JESSIE COLE

DARKNESS ON THE EDGE OF TOWN

'This work operates at deeply engaging and emotional levels while excellent story-telling drives it.'

—Nigel Krauth, *Westerly*

'*Darkness on the Edge of Town* proves difficult to put down as it hurtles towards its confronting conclusion.'

—*Who Weekly*

'... it's an accomplished portrayal of how seemingly random events can trigger life-changing outcomes.'

—Sunday *Canberra Times*

'Jessie Cole's spellbinding first novel is the kind of book that you can describe with words such as "beautiful", "touching" and "tender" as easily as you can with words like "uncomfortable", "painful" and "disturbing" ... I can't wait to see where this talented new voice takes us next.'

—*Good Reading Magazine.*

'A sad and tender tale of the extraordinary events which make up the everyday lives of ordinary people, *Darkness on the Edge of Town* elegantly expresses the simplicity of emotions that we often find so hard to handle. Unflinching in her capacity to scrape at the raw nerves of our desperation for love, Jessie Cole has written a distinctly Australian story about hope, desire, sexuality, violence and our failure to communicate.'

—*ABC Radio Brisbane*

For R.

The steering in the old girl lunges a little to the left, so on that night I was holding tight around the corners, swinging into them the way Marie says she hates. She can just see herself plummeting down the drop on the side of this mountain, but I've lived here for years and I know the road pretty good. It's real green and bushy out here, and in the night it can look like there's nothing, no houses, just this winding precarious stretch of road. It was late, and I'd dropped Gemma off earlier at her girlfriend's house. She's sixteen, my girl, and she's only just reached that girly stage. Nail polish and lip gloss. She came home from school the other day all dolled up. It was photo day and her friends had taken her aside and done her makeup. I reckon she expected me to hit the roof, to tell her to 'get that shit off', but I just looked and didn't say nothing. She washed it off anyway, soon as she got home.

That night, I was on my way home from the pub. I don't drink much, just a couple of beers, but I like to see the boys now and then. It was coming into winter and the air inside the truck was cold. I lit a cigarette, banged around the final bend before my house, and right there, right out front, was the upturned car, engine still running. The lights of the car were shining down into the bush, lighting up the dark. I pulled up on the grass out front of my house and ran over to the car, peering in the window, but there was no-one inside. The smell of exhaust fumes lingered in the air and, reaching in, I turned the engine off. There was a flicker of movement on the side of the road. She was squatting there, swaying slightly, the bank dropping away steeply behind her. Humming—she was humming. In the moonlight she looked kind of crumpled and broken, her long dark hair falling forward over her body.

'Shit, honey, you right?' I asked, but she didn't move, as though she didn't hear, and so I crept up a bit closer. 'Mate, you okay?'

She looked up then and her hair fell away, and I could see in the shadow of her arm she held a baby. Its body was limp, its eyes closed.

'We got to get you some help,' I said, and she whimpered. I crouched in front of her, reaching out a hand. 'Sweetheart, you need some help. Come on off the road.'

I wondered how long she'd been there, perched like that on the edge of the drop. I needed to call an ambulance but I didn't know if I should leave her. Reaching out two fingers, I tried to feel the baby's pulse. The baby was cold and I couldn't feel anything. I didn't know CPR anyway. Up close I could see her. She had eyes like an animal caught in a trap, large and kind of misted, dark. The side of her face was bruised. She was youngish, I reckoned, late teens or early twenties. Not much older than Gemma. Didn't look like she was from round here. Looked foreign, sort of. She held the baby tightly with one arm, the other hung dangling from her shoulder. Her shirt was lifted and her pale breast sat exposed above the baby's head, dripping milk. Watery white drops that plonked slowly on the baby's slack face.

'Oh, honey, come on, come on off the road,' I said, starting to panic.

A line of blood crossed her face and fell in a sticky blob on the baby's foot. She shuddered and toppled backwards, landing with a jolting thump, the baby's head flopping sideways. Dusty pebbles rained down the slope behind her.

'Love. See that house there? That's my house, and I'm going to pick you up and take you inside so I can call an ambulance, okay?'

I didn't know if I should move her, but I figured if I ran inside and called an ambulance and she went over the edge, that was worse, and I wanted to check out her head, see if I could stop the bleeding.

She didn't try to move away, but I guess she was pinned, holding the baby in one arm, and her other hanging loose like that. I stepped alongside her, close up to her baby's head. It looked a strange kind of blue in the moonlight, and I felt suddenly sick. Slipping my arm beneath her legs, I scooped her up. She clung to the baby and sort of curled up around it, her loose arm swinging out and falling back against her. If it hurt, she didn't make a sound.

I shoved the back door open with my knee, staggered inside and sat her down on the piles of unsorted laundry on the couch and then made the call. The emergency services said, 'Ten minutes,' and 'Try to stop the bleeding.' I couldn't bring myself to say the baby was dead. When the woman asked me, I just said, 'Not good, not good.'

'Is it breathing? Pulse?'

'Nah, I don't think,' I said, watching the girl's face as she sat dripping blood on my laundry.

'Try and clear the baby's airway.' The woman's voice was matter-of-fact.

'Okay,' I said, and hung up.

'What's your name?' I asked the girl, a towel in my hand.

She looked up at me, the stripe of blood beside her eyes, but she didn't speak. Her lips were dry and her breast still dripped milk. A large wet patch had spread across her shirt on the other side.

'Can we wrap it up?' I asked, holding up a towel. 'It looks cold.'

Her eyes narrowed, and I leant down to put the towel around the baby. 'I'm just going to clear the airway.'

I hesitated a few seconds before I put my finger in the baby's mouth. I couldn't feel anything inside, though I didn't know what I was looking for. The sight of my finger inside that baby's mouth felt wrong, like a violation, and I wished I hadn't done it.

'It's a he,' she whispered, and then the tears spilled from her eyes in great long lines, though she stayed still, almost unbreathing.

'I'm going to look at your head, okay?'

I moved her hair aside, looking for the source of the blood. Her hair was sticky and matted, black. I mopped up some of the blood with a washer, and then I saw the gash. It was smallish, just a little way into her hairline, and I felt the air rush out of me in a sigh. 'It's alright, just a scratch really. You always bleed a lot with a nick on the head.'

My fingers trembled as I held the washer over the cut. Through her hair I could see the baby's face. It looked sunken and I had to turn away.

'They'll be here soon.' I wasn't sure what else to say.

When I glanced down at her, the pearly whiteness of her breast seemed to gaze at me like an unblinking eye. I realised she couldn't cover herself even if she wanted to, with the baby in one arm and the other bunged up. I pulled the washer away and saw that the blood had stopped. Kneeling down in front of her I tried not to glance at her dripping breast.

'Do you want me to pull your shirt down for you?' I asked, keeping my gaze on her face. She looked down through her wet eyes and pulled the baby in towards her. Holding it against her breast, she shook her head. I could see her shoulders begin to shudder, and then she breathed out real hard, in a kind of burst. She moaned deep, and the room was suddenly unbearably small.

I knew I couldn't just crouch there and watch her so I picked her up like I had before. Pushing some of the clothes aside with my foot, I sat back down on the couch with her on my lap. Holding her tightly, as she held the baby, I tried to absorb some of her shaking inside me. She moaned again, and I pressed her in against me. She quietened, and I don't know why but I rocked her gently, like she was just a baby herself, and she whimpered against my chest. Then I heard the siren.

They knocked on the door and I called out. The two ambos came in with their orange cases, and I felt the girl tighten and curl up smaller on my lap.

'Where's the baby?' the woman ambo said, glancing sideways at the man, then at me.

'She's got it here, it's wrapped up in a towel.' I motioned downwards with my head.

The woman knelt in front of us, and carefully pulled back the girl's hair. 'We've come to help you, but you need to let go and show us the baby.' Her voice was calm and kind, but somehow routine.

'You the father? How old's the baby?' the man asked. 'You in the car? You injured too?'

I held up my palms. 'I don't even know the girl. I just came home and found her on the road. I don't know nothing.'

He looked at me, with her curled up on my lap, but all he said was, 'Right.'

'Any ID on her?' the woman asked.

'Dunno, mate, I haven't checked.'

'Her arm?' The ambo leant over his work mate's shoulder to look at the girl.

'Yeah, I'd say she's broken the humerus,' the woman answered, reaching in to try and unwrap the baby. She searched for the pulse and breath, and then put her fingers inside his mouth like I had. The man ran outside and I could hear the clunking of the trolley as he got it out of the ambulance.

Fingers on the baby's neck, the woman ambo looked up at me.

'Were they still in the car when you got there?'

'Nah.' I shook my head. 'She was right on the edge of the road, looking wobbly, so I thought I'd better bring her in.'

'Can you get her to turn towards us?' the woman asked me. 'We need to get at the baby. We need to start CPR.'

I lifted the girl a little and tried to turn her outwards, but she shifted in deeper towards me, pulling her arm tightly around the baby.

'Come on, mate, they're going to try to help the little fella. Do some CPR,' I whispered in her ear. 'Let go, darling, give him over.'

She wouldn't move. When the ambo was back inside, together they took the baby from her. Whimpering, she tried to resist them, burrowing in against me. In the end the man prised her arm away so the woman could grab the baby. I tried to soothe her, but I could feel my voice go all hoarse and wet. Once they had him, she seemed to sag inwards. Started up shaking again. I could hear her teeth chattering, and the man threw me a big white blanket while the woman held a plastic mask over the baby's mouth.

'She's in shock,' the woman said. 'Another ambulance is coming for her. Just hold tight for a few minutes while we work on bub.'

I wrapped her up in the clean-smelling blanket, holding her while they tried to revive the baby. The woman ambo held the mask, while the man pressed his fingers gently against the baby's chest at a weirdly fast pace. It looked pitiful, how softly he pressed, and I closed my eyes, not wanting to watch.

'Adrenaline?' the woman said, and the man just grunted out a 'Yeah.'

I opened my eyes. They jabbed in a needle and connected the baby up to a bag of fluids, but nothing seemed to happen. I could see a spot of his mother's red blood on his tiny foot. The other ambulance arrived then, and they carried the baby out on the stretcher.

As she left, the woman ambo turned and said, 'Tell the mum we're taking him up to the hospital, and she'll follow in the other ambulance. Mum? Did she hear that?'

The girl didn't stir in my lap, just shuddered a little, and I nodded my head.

The second ambos came in, two men with their trolley, and started working on her. 'What's your name, love?'

She looked away.

'What's her name, mate?' the man asked me.

'I dunno, I just found her. I don't know her.'

They took her pulse, temperature and blood pressure, and kept on talking to her, soft and insistent, but she didn't respond. One man reached over and gently pulled down her shirt to cover her breast. The other shone the torch in her eyes and carefully put her arm in a sling.

'How is she?' I asked, starting to feel a bit shell-shocked, I spose.

'The arm's definitely broken but we don't think there are any internal injuries. Bit bruised up, but I'd say she's lucky. She'll be fine.'

I thought of the limp baby on the trolley, but I didn't say nothing.

'Now let's see if we can get her off your lap and onto the stretcher.'

She whimpered, clinging to me with her good arm, and they had to unfold her fingers one by one from my shirt. As they sat her on the trolley, she looked at me with those large dark eyes and then she shook her head. The ambos put a neck brace on her and tried to get her to lie down. Staring at me, she clutched the trolley with her good hand until her knuckles went white. I looked back at her from my spot on the couch, and then I couldn't help it, I covered my eyes with my hands.

'We're taking her up now, mate,' the ambo said, patting my shoulder. 'You did good.'

Rubbing my face with shaky hands, I listened to the sound of the trolley clunking up into the vehicle. I couldn't stop thinking of the dripping milk and the baby's blue face, and the way she looked at me, eyes spilling fresh tears, as they wheeled her out the door.

2

The bus moved up the mountain real slow, like every other day, and by the end I was the last kid still riding. Not many people live up here on the ridge line—well, not many kids, anyway. My best friend, Mel, lives in town, so she gets off right at the start, and then I'm left just staring out the window.

When I walked in there was no-one home. The house was quiet and still. Dad always looks pooped when he gets back from work, so I try to make dinner most nights. Pouring a glass of cordial, I started checking out what we had in the fridge. It was covered in stickers, our fridge, and not ones we'd put on. They were an ugly bunch of pictures, peeling down at the corners, mostly trying to sell stuff that probably doesn't even exist anymore. I tried to stick down the edges once so they'd be flat, bought some glue from the shops and everything, but it didn't work, they just curled back up again. The door's lost suction, but it's got this piece of Velcro glued onto the side that holds it shut.

I'd forgotten to get the sausages out of the freezer before school, so I put them in the sink in hot water and hoped they'd defrost before dinner. There were some potatoes in the bucket under the sink, so that would have to do. Sausages and mash, a can of peas to top it off. Marie and the boys wouldn't expect anything fancy.

Dave was watching me that afternoon when I got on the bus, I could feel his gaze on my back. He stood with a bunch of boys along the wire fence outside the school. I wanted to turn around and wave but I didn't. It seemed strange that in a few hours he'd be at my house with Marie, but at school he pretended he didn't know me.

I drank the last of my cordial and started peeling the potatoes. The first peel came away in one big spiralling curl, and I held it up to the window and jiggled it a bit. I liked the way it hung from my fingers and bounced in the air. I stopped myself then, cause it was only a potato peel, and I imagined what Dave would think if he walked in and saw me doing something so loopy.

When I thought of Dave and his gaze on my back, something happened to me. A kind of prickling sensation spread up the back of my legs. I wondered what it was. It felt like something secret. And just for that minute I was glad that Dad wasn't at home to see.

My dad, he collects broken things. We live near the dump— not near enough that you can smell it or anything, but close enough that on the way home from work Dad can drop by and pick something up. Where other people see junk he sees potential, and our house is full of things that don't match, or that lean slightly to one side. Laminex chairs, all differ- ent heights, around a banged-up wooden table. Sloping old bookshelves, cupboards with missing doors. Even the oven's a ring-in, propped up against one of the walls on fat wooden legs. Dad says it's probably a fire hazard, but the house hasn't burnt down yet. I've heard him brag that he has never paid money for furniture, and I can vouch that it's true.

I don't mind all the broken things—sometimes I shift a chair outside when I think the house is overflowing, or when I can't get to a kitchen cupboard or something—it's the people that bother me. My dad collects broken people too. In a roundabout way that's how he got me. He met my mum when I was about two, and he married her. She's my mum so I don't like to say bad things about her, but she is pretty full-on. When Mum ran off, Dad was left with me, but she still comes back now and then, looking for things to pawn. Maybe that's why Dad keeps the house full of junk—so there's nothing worth stealing. I used to think that everybody was as crazy as my mum. Now I realise there are plenty of normal people out there.

But my dad, he's had this string of crazy women since my mum. Some of them are probably nice enough, but they've all had that edge. He doesn't pick them up in pubs, nothing like that. He meets them down at the sports field. He trains a group of boys for the footy club, and the mums hang around to watch. They're a faded bunch, usually, as though their kids have sucked the life from them. They'll see he's a nice bloke, and they know he's a single dad. They notice him talking to the kids, and playing with them, even the littlest ones that have just come to wait for their big brothers. He'll have them all laughing, and those mums, they'll start to watch him instead of their boys.

In the end they'll hang around after training, waiting for their turn for a chat, and he'll smoke a cigarette and listen, and sooner or later their whole sorry life story will come out. One mum will be particularly persistent, or her story will be longer or sadder, and Dad will give her a bit of extra time. And while they talk, her kids will race around him, competing with their mum for a piece of his smile. He'll rough around with the kids, tell them jokes, talk a million miles an hour like he does, and the woman will start to look lively again. Like she's been parched, and it has suddenly rained. When the sun's going down and she can't possibly stall any longer, she'll leave, and my dad will watch her go, and I'll see that she's become somehow different in his eyes. Next training session I'll watch him watch out for her, and she'll come early, and then I know it'll start up soon.

At home he'll be acting all cheerful, whistling in the morning before work, and then soon enough the woman will drop by with a bunch of little kids. The two of them will chat outside and have a smoke and a cup of tea, and maybe she'll stay for dinner, but she'll always have that edge, that half-crazy edge. The woman will be wired and talk too much, and stand too close, and laugh too loud. And the two of them will go along happily like that for a little while, and then it will be the same as the last. Something will come up, some everyday drama, and the woman will go nutty. Dad will try and stick in there, help her out, whatever, but she'll start dropping round at all hours, acting forceful and intense, or the ex-husband will. Dad always thinks he can sort it out—keep everything in check—but things get messy, and then the shit will hit the fan again.

This latest woman he's met, Dave's mum, Marie, she's alright in some ways but she's insanely jealous. She thinks Dad is a ladies' man, and every time she comes round they end up fighting about some woman on the street who looked at Dad sideways. Since as far as I've seen Dad never makes a special effort to pick up women—they always seem to trip over each other to fall in his lap—it makes her come across as a little crazy.

Dad says she's just 'insecure', married to the same man for twenty years and now she's out on her own and doesn't feel like anyone could love her, but sometimes I wonder exactly why he does love her. She's not pretty, not like the porno girls in the magazines under his bed. Dad's a few years shy of forty but she's a bit older than him, kind of rugged-looking. You can tell she doesn't feel like an attractive person. When she talks, she bunches up her face and looks mad at the world. She just seems full of rage, and after a few drinks, you know, it will all be on.

I used to wonder what it was about my dad that attracts these women, how they must be so desperate for a listening ear. But lately I'm thinking maybe I should be worried about what it is that he needs from them.

3

I'd just got back from training the boys. Marie had come round with her two young blokes, and we were settling down outside with a beer before dinner. Gemma was inside cooking, which was great since I'd been up real early and I was buggered. She's like the woman of the house now, my Gem, though she doesn't sort out the laundry. Marie's boys were kicking the footy behind us in the backyard, and I looked at the sky and saw the moon was already up, though it wasn't dark yet. I stood up and stretched, sore from using the chainsaw. I work in this old bloke's orchard, maintenance stuff, and it's alright, but sometimes I come home pretty tired.

I went to hang a piss in the garden, and saw her out on the road—the girl from the crash. Fumbling with my pants, I stepped down the driveway towards her. She was watching the ground, just standing there. She looked a mite unsteady. Her arm was plastered all the way up to the shoulder and she held her free hand beneath it as though cradling its weight. The unused sling hung limply round her neck like a colourless bandana.

'Hey,' I said softly. I didn't want to scare her, but she looked up startled. Right away I saw that without all the blood she was a real pretty girl, exotic-like, and I knew Marie was going to go nuts. The side of her face was yellowish with bruising. She let go of her plastered arm and tucked her hair behind her ear and I realised that I'd been staring.

'Where's the car?' she said, looking at me straight.

'Got towed the next morning. Into town I guess,' I said, and the weight of the plaster seemed to drag her to one side. 'The police came and took some photos, asked me a few questions. Couldn't tell them much. It's a bad fucking corner. They organised the towing.' I wanted to reach out and hold the arm for her but I didn't. 'That was a few days back. How you holding up?'

She didn't reply to that, but I could see a fine sheen of sweat against her forehead, and her hand trembled as she pulled a long piece of her dark hair forward and rubbed it against her lips.

'How did you get here, mate?' I asked, looking around for a car.

'Hitchhiked, from town.' She wiped the back of her hand against her eyes. When she looked at me again I could see she was in pain. 'Had to walk the last bit, no cars.'

'Fuck, you don't look like you should be out of hospital yet, love.' I stepped closer to her then. 'Do you want me to give you a lift back?'

She shook her head, hard, and two tears popped from her eyes.

'I'm not going back. You can't make me go back.'

I put my arm around her, and she leant into me.

'I pulled the IV out, cause they wouldn't let me go. I had no clothes so I took the lady in the next bed's stuff when she was in the shower. It took forever to get dressed in the toilet with my arm. I had to get out. I hope they weren't her favourite things.'

The words tumbled from her mouth. I looked down at the clothes she was wearing. They were real big on her, the shirt sort of old-fashioned and the pants hanging down over her feet. I could see her bare toes sticking out.

'You don't have any shoes,' I said, imagining her trudging along the winding road to my place. I had to take the girl inside, but I was thinking of Marie and the shitfight I'd have on my hands, so I just stood there, holding her up and wondering what to do.

'Yeah, the lady's shoes were too big.'

There were blood spots on the shirt from where the IV had been. I pulled the sleeve up to look underneath. Her arm trembled, the skin translucent beneath my fingers, blue veins showing through. The small hole from the IV was puffy and raw.

'How'd you pull it out with only one hand?' I said, remembering how my wife used to stick needles in her veins when she thought I wasn't around.

'Wrapped it round the trolley thing then pulled.'

She sagged against me and I tightened my grip on her.

'Vinnie, what ya doing out there?' Marie had come looking for me.

'Well, who the fuck is this?' she said, stomping towards us. I could see her looking at my fingers on the girl's arm, and I held up my hand, trying to slow her down.

'Marie, this is the lady I told you about, who crashed the car outside the house.'

She stood still, looking at us.

'Well, you two look close. I thought you said you didn't know her.'

The girl trembled in my arms. Her head was hot and limp against my shoulder.

'I don't know her, but she's not well. She's hurt.'

Marie took a sip of her beer.

'What the fuck's she doing *here*?'

I just held up my palm again, in surrender. What did I know? But I started walking the girl inside.

'She hitched here from the hospital,' I said, willing Marie to be calm. 'What do you want me to do? Leave her out on the road?'

Leading her round the back, I sat her down in one of the old unravelling cane chairs.

'Gem, you want to bring us a pillow?' I called through to the kitchen, and Gemma went into the lounge and came back with a cushion. My daughter looked at the girl, in that quiet way of hers, and then she looked at Marie and I could see she was worried that Marie was going to blow. I put the cushion under the girl's plastered arm to try and hold some of the weight, and Marie's boys stopped kicking their ball and came over to have a look.

'Do you want a beer?' I asked the girl, and she nodded, but she didn't really look at me. In the doorway I unscrewed the top and leant out, handing it to her. She sat in the chair, her eyes kind of glazed, and no-one said anything for a bit. After a minute the girl took a sip, and then the tears came. Silent and wet, they slid down her cheeks in tiny lines. Women crying always make me edgy, like I need to do something to make it stop, but Marie's boys just stood there and gawked at her.

'Aw, Vince, she's beautiful, man, who is she? Where do you think she's from?' Pete said, as though she wasn't even there. Dave, the older one, nudged him in the ribs and tried not to smirk.

Marie's body tensed, fingers clenching her beer. I shook my head at the boys and they stopped staring and looked at their feet instead. Gemma was hovering in the kitchen, half-way through cooking, not knowing where to be. I fidgeted with a cigarette, lighting it up and sucking in a deep lungful of smoke. Marie watched me, her eyes narrow and hard, and the girl's tears just kept running down her face. Her sorrow seemed to spread out from her like water. If Marie wasn't there I would have gone to her and tried to do something, I don't know what. But Marie looked about to explode, and dinner was nearly ready, and I felt kind of stymied there in the doorway.

Finally, the girl glanced up at me. 'I think I'm drowning,'

she said, then she stood up and walked over towards the trees.

Pete looked freaked. 'Vince, what's she crying for? What's going on?' he said, dropping the ball.

'She crashed her car out the front a few days back and her baby died. I found her on the road. I don't know what she's doing here.' I turned towards Marie. 'Her baby died, Marie. I helped her just like you would have if she crashed in front of your house.' I stepped out of the doorway. 'Listen, I think I have to bring her inside. I don't think she should be walking around. It's not just that arm—she copped a gash to the head. I'll set her up on the couch and then we can all have dinner.' Putting my beer down, I stubbed out my cigarette in the jar lid on the outside bench. 'Okay, Marie?'

Marie stood up, swirled the beer around in the bottle and then took a final swig. 'Well, we aren't hanging around waiting for you to work this one out, Vinnie. Come on, boys, let's go.'

Dave and Pete shook their heads, looking from Marie to me, but after a few seconds they started heading slowly for the car. Dave leant down to pick up the footy on the way, and I could hear him mutter some swearword beneath his breath.

'Marie, this is fucked up. Gem's cooked you dinner,' I said, even as I moved away from her towards the girl in the trees. 'Just stay, come on.'

Marie looked over at the girl, sagging now against a tree trunk.

'Fuck, Vinnie, I just can't deal with it. You and your shit.

Get it sorted, then call me, right?'

She called out a low goodbye to Gem and then stalked out to her car. The boys were already waiting, and when she reversed, they stretched their arms out the windows and waved.

When I reached the girl, she was leaning her head on a tree, pushing it hard against the bark. I thought of the gash on her head, and the long stripe of blood that had splashed her baby's toes, and I eased her back, away from the tree. With her good arm she clutched her breasts, and I could see the milk spreading out beneath her fingers and across her shirt. She sobbed then, all broken and clogged, and her knees seemed to buckle beneath her.

'Come on, mate, let's go inside,' I coaxed and half dragged her towards the house.

Gem stood in the doorway, dinner laid out on the table. Dinner for three.

'Gem, help me clear off the couch.' I lifted the girl in the door and Gem hoisted up bundles of clothes and threw them into her room. 'Okay, now get some more pillows, let's prop her up.' I held the girl sitting while Gem arranged the pillows behind her. She groaned and closed her eyes, still holding her breasts. Her plastered arm looked stiff and heavy against her belly.

'What's wrong with her, Dad? Why's she all wet?'

Glancing at Gem, I saw she was scared. 'The bub was only real little, so now her breasts are full of milk, but he's not here to drink it, I guess. Gem, I think she's got a fever. I think I should take her back to Emergency.'

The girl on the couch scrambled up, wide-eyed and wild-looking.

'I'm not going back. If you want me to leave I will, but I'm not going back.' She looked about ready to bolt out the door.

Reaching out, I touched her forehead. It was hot and clammy. 'Mate, you're not well. You're sick, and I don't know what to do. I think you need a doctor.'

She shook her head. 'I'm not going back to the hospital. You can't make me.'

I watched her a minute and she started to shake a little. 'What about a doctor in the morning, then?' I asked, thinking maybe I'd give her a Panadol and see if she'd go to sleep.

The girl nodded, slowly, and I helped her lie back down.

'Okay, a doctor. Not the hospital. In the morning.' Motioning to Gem to sit down at the table and eat her dinner, I went and got a Panadol and helped the girl sit up to swallow it.

'Just try to go to sleep, okay?' I said, pulling the sling over her head, and covering her body in a blanket. She looked real tired lying there, and I moved away, sitting down with Gem to eat my dinner.

The girl slept on the couch, whimpering a little now and again, and after Gem went to bed I rang my sister, Jan. She's a nurse and she's had a few kids, and I thought she might know how to help with the leaking milk.

'Fuck, Vinnie, let me get this straight. A couple of days ago you found a girl with a dead baby on the road outside your house, and now she's asleep on your couch with leaking tits, and probably mastitis and you want to know what to do?'

'Yeah,' I said. 'I spose it sounds a bit weird.'

'You're like some kind of magnet for disaster, aren't you?'

'What should I have done?' I heard myself growl into the phone. I didn't know how I could have acted differently. Glancing over at the girl, I saw her shift in her sleep and hoped I hadn't woken her.

'Take her to hospital,' my sister said with a sigh.

'She's shit-scared of the hospital, I don't know why. She pulled out the IV to get out of there. I'll take her to the doctor tomorrow, okay? But is there anything I can do about the milk? When they leak she can't do nothing but cry.'

'What a fucking story, Vinnie.' Jan sighed again. 'Okay, do you have a cabbage?'

'A cabbage? What for?' I shifted the phone to my other ear.

'Do you have one?'

'Probably, in the bottom of the fridge.' I wandered over to the fridge to check. 'Yep. One cabbage.'

'Okay, what they usually say is, run a hot bath, and once she's in the water, squeeze out as much milk as you can, sort of massaging out any lumps. Then wrap the breasts up in cabbage leaves—you know, inside the bra—and change the leaves every few hours.'

'Are you for fucking real?' I said, stunned. 'I can't do that.

I don't even know her. I can't fucking massage her breasts.'

'You asked me a question, I'm just giving you the answer. Sounds like an old wives' tale, but that's what they do in the maternity ward in town. If the mum gets all engorged—bath and then cabbage leaves.' My sister's voice was firm. 'I told you to take her to hospital.'

'Alright, well, I'll just wait till the morning then. Hey, thanks Jan.' I was ready to hang up.

'Anytime, Vinnie. Ring me back and tell me how it goes. Mastitis can be real painful. She'll feel like she's been run over by a bus. It's an infection—make sure you take her to the doctor in the morning.'

'Yep. Hey Jan, don't tell Mum, alright?'

'Vinnie, if you got a woman shacked up there with you, Mum's going to find out in no time, you know that,' my sister said with a sort of snorting laugh. 'Have you told Marie?'

'She was here this arvo.'

'Bet she's taking it well.'

'Yeah, great. Well, see ya, Jan.'

I hung up the phone and went over to look at the girl. She was a bit pink-cheeked. I touched her head and it was still hot.

'Fuck, this is a fuck-up,' I said to myself, and headed off to bed.

4

I woke with a start. I could hear the girl moaning out on the couch. Shaking my head I got up, put on a shirt and walked out, switching on the light. The wall clock said 2 am. The girl was flushed, and she squirmed under the sudden brightness. Her whole shirt was soaked through, and the blanket wrapped around her was wet against her breasts. I could smell the sweetness of the milk. Opening her eyes, she blinked, cupping her breasts in her left arm, a low humming groan coming from the back of her throat.

'Listen, I got a cabbage in my fridge. You heard the thing about the cabbage?'

'What?' she said, struggling to sit up, her eyes kind of filmy. She looked half mad, feverish and shaky.

'I rang my sister. She said you need to get in a bath and squeeze out the lumps, and then wrap them in cabbage leaves,' I said, pointing to her wet chest.

'They hurt so much, they throb.' Her voice was rasping and low.

'You want to hop in a bath?' I asked. 'My bath's pretty ancient, but it'll do the job.'

'Okay,' she muttered, and I hoped she knew what she was saying. Walking into the bathroom, I turned on the taps. I could hear her struggling with her clothes in the next room, and then she started to cry. I came out from the bathroom, and she was standing in the centre of the room, swiping at her eyes with her free arm.

'I can't do it,' she sobbed. 'I can't do it one-handed.'

'Let me help. Come in the bathroom and I'll help you.' I moved aside for her to go in. 'I'd better get a plastic bag for your arm.'

She sat on the edge of the bath, a bit wobbly, and I went off to get the bag.

'Okay,' I said when I came back, 'now we're going to do this real slow. First we'll pull your good arm through, then your head, then the plaster.' I lifted her shirt and she manoeuvred her left arm out of the sleeve. The shirt made sucking sounds as it came away from her body. I pulled it carefully over her head and then over the plastered arm. The fabric was heavy and wet in my hands and I chucked it into the laundry bucket. She wasn't wearing a bra, and with her good arm she covered her swollen breasts, but I could still see how red and sore they looked. The skin seemed tight and angry, and the milk dripped slowly over her forearm and through her fingers. She had a curve of small oval bruises on her upper arm and a mark on her neck from where the seatbelt must have cut in.

Leaning over the bath, I turned off the tap.

'Okay, I'm going to put this plastic bag over the plaster, with a rubber band at the top, but you're still going to have to keep it from getting wet.'

The plaster was a bit wet and sticky from all the milk. I pulled the bag up her arm, then the rubber band, and she stood up to take off her pants. One-handed she tussled with the button, and in the end I leant down and just pulled the pants off. The darkness of her pubic hair against her skin surprised me, and I stepped back real quick, not knowing where to look.

'I've got no undies. I couldn't steal some other lady's.' She was shivering all over now, though the bathroom was warm with steam. 'They gave me something and when I woke up I was on the ward in a hospital gown. All my clothes were gone. No-one could tell me where.'

'Come on, in the bath. Can you hold your arm out of the water?'

I tried not to look at her nakedness while I helped her in. She couldn't lift the plastered arm, so I sat on the edge of the bath and held it out of the water. Her breasts seemed to float on top of the water, and I wondered if the whole breast-massaging thing would work if they weren't submerged. She couldn't lie down lower without wetting the plaster. Leaning towards the bathroom cupboard I pulled out a washer. Still holding up her arm, I dipped it in the bath and laid it over her breasts.

'Do you know how to get the milk out? Massage the lumps?' I asked, sort of stumbling on the words.

'Yeah, I think. I expressed milk once into a bottle.' She

pulled the washer aside and tentatively touched one breast. 'It's just so sore.'

'Fuck, I should have given you another Panadol. I didn't think.'

Stuck there holding her arm out of the water, I couldn't go and get the Panadol. I tried not to watch her fingers moving against her breast, but no matter where I looked I could see the movement of her hands from the corner of my eye. Suddenly I felt like I was in some kind of surreal porno film, and as soon as I thought it, I went hard. I was wearing a pair of tracky pants, so I just hoped to Christ she wouldn't notice. She sucked in a sharp breath, squeezing her breast between her fingers, her face shuddering with pain. Her expression was so close to pleasure, I couldn't help it—I watched as milk sprayed out from her dark nipple like a shower and melted away into the bathwater. Her face was red from the hot water, and the long ends of her hair floated out around her body. She looked like some kind of beautiful maimed mermaid. Tears dripped out from under her closed eyelids, but she kept squeezing at the breast, and the milk kept flowing. After a few minutes, I shifted around, swapping arms, cause the plaster was heavy. When she opened her eyes and looked at me, there was such a starkness there, I wanted to look away.

'Now the other one, come on, you're doing good,' I said, hoping she wouldn't look up and see the tent pole in my pants. She lifted her hand and pressed it hard against her forehead. I could see she was trying not to sob. Pushing her knuckles against her eyebrow, she moved the washer over and started on the next breast. This time she was quicker, and I watched as the milk streamed out. I don't know if it was my imagination but the water looked soapy with milk.

'Is that enough? Surely that's enough.' Her voice was wet sounding, and the steam had formed balls of water on her pink face.

'Yep. I hope so.'

Still holding up her arm, I turned around to reach for a towel. When she stepped out of the bath, she clung to me with her good hand and gritted her teeth, and I could see her knees were giving way.

'Okay, hold on,' I whispered, sitting her down on the edge of the bath. 'I'll wrap you up and carry you out, right?'

I took her into my room. That's where the clothes were.

'Just sit here.' I put her down on the bed. 'I'll get the cabbage and the Panadol.'

When I came back she was staring round my room, a little dazed. I turned on a lamp in the corner and switched off the overhead light.

'Bit rough, isn't it?' I said, looking at the half-finished paint job on the wall, all my stuff stacked haphazardly around in piles, a mess. 'The people who lived here before, this was the kids' room, and they'd painted a mural on the wall, but it was kind of spooky. All these *Alice in Wonderland* characters. So after a few years I painted over them.'

'Why didn't you finish the wall?' The tops of her shoulders

were pearly and smooth, poking out from her towel.

'Ran out of paint or something. I dunno.' I gave her the Panadol and a glass of water, and then opened up my wardrobe and had a look.

'Okay, I reckon our best bet would be trackies and a big jumper, with one of Gem's little singlets underneath to hold in the cabbage.'

I propped the cabbage under my arm like a ball and crept into Gem's room to steal a singlet. I could hear my daughter's sleeping breaths as I opened the door. All the clean laundry was still piled on the floor, so at least that part was easy.

Back in my room I pulled out some trackies. The girl stood up, still wobbly, and I helped her step into the legs. I guess she was damp, cause the material sort of stuck. I had to lean down and pull her feet through. She gripped hold of my shoulder and I felt my cheek graze her belly. I had no idea what I was supposed to do with the cabbage, but I thought we'd better get the singlet on first. Pulling the plastic bag off her plastered arm, I looped the armhole of the singlet over the plaster. I pulled the singlet over her head, and she pushed her good arm through. The singlet was stretchy and a little tight, and I figured it was a bit like a bra.

'Okay, now the cabbage.' I pulled the outside leaves off the cabbage and threw them on the cluttered desk. The next layer looked clean enough so I tore those leaves off and handed them to her.

'Can you pull the singlet down for me?' she said, and kind

of smiled, like it must have just dawned on her how odd it was to be wrapping up her breasts in cabbage with a strange man at three in the morning. I'd never seen her smile. Her whole face transformed, and she didn't look quite so beautiful. She had a little gap between her front teeth which made her seem suddenly cheeky. Her eyes were large and glassy, feverish.

I reached out and pulled the singlet down on one side, and her breast sat softly against my knuckles. It didn't look so swollen and tight, though it was still a bit red in patches. I went hard again, just watching, but she didn't notice cause she was concentrating on wrapping up the breast one-handed. She finally got it, and I quickly pulled up the singlet, then we did the other side. The smell of cabbage filled the room and she wrinkled her nose. I went out and put the rest of the cabbage back in the fridge, and then I came back in and helped her with the jumper. She sat down on my bed, all trace of the smile gone.

'Please don't make me go out there,' she said, looking at the floor.

I watched her, all bedraggled and broken on my bed, but I didn't say anything. I thought of Marie's angry stalking footsteps as she stomped off to the car.

'I just mean, I don't want to be by myself.' Her voice wavered and cracked. Her long black hair was still tucked up in the back of her jumper, and I realised maybe she couldn't get it out. Setting one knee on the bed, I pulled the hair out from behind. It was wet and tangled.

'Sure. No skin off my back. I'm told I snore real loud though, so be warned.'

She lay down then, and I lay down beside her. I leant over to switch off the light, pulling the doona up over us both.

'Hey, tomorrow you better start wearing that sling. That arm is fucking heavy,' I said into the darkness.

'Yeah, okay.'

I could feel her breath against my cheek.

'And mate, tomorrow maybe you can tell me your name.'

I could feel myself drifting already towards sleep. Mostly I'm like that, out like a light when my head hits the pillow. She said something quiet and low, a murmur. I just caught a wisp of her words, but I didn't grasp them, and then I was gone, swallowed up by sleep.

5

I woke early cause I'm usually at work by seven. Soon as my eyes opened I knew someone was watching me, and sure enough Gem was standing in the open door, just staring in, her face tight and closed. I felt the weight of the girl against my body then, her head cradled in my arm, her leg draped over mine. I eased myself out from under her. It was still pretty dark and I moved quietly from the room and closed the door behind me.

'You're up early.' My voice was rough. I needed a coffee. Gemma didn't say anything, just walked away from me into the kitchen.

'She got scared. Didn't want to be by herself,' I said, following her out and sticking my head under the kitchen tap to wake up. 'She's in a mess, what can I do?'

Gemma got herself a bowl and tipped in some cereal. She didn't answer me, but when she opened the fridge door to get the milk it swung backwards hard and hit the wall. She looked surprised by the sound, by the fact that she'd made it, and grabbing the milk she quickly closed the door. When she sat down at the table, her eyes wouldn't meet mine.

'Gem?' I wiped my dripping face with a tea towel.

'Dad?' she said in reply, looking out the window at the

sun coming up.

'If you got something to say, say it.'

I watched my daughter spoon cereal into her mouth, looking anywhere but at me. She looks just like her mother did, Gem does. Tall and gangly like a foal, all legs and arms. Light-haired. Real pretty. She'll be taller than me soon, I reckon.

'How long is she going to be here?' My daughter's voice was low but steady.

'I don't know. I'll ask the doctor what to do. Call in Community Health or something. She must have family somewhere.'

'Or a boyfriend,' Gem said, and looked me straight in the eye.

'Yeah, or a boyfriend,' I agreed, turning away to put some toast in the toaster. I poured Gemma a glass of orange juice, and sat it on the table. 'Gem, nothing happened, I promise.'

'No? I got up to go to the toilet, saw her pants on the floor.'

'She woke up covered in milk, I helped her get changed.'

Gem picked up the glass and sipped her orange juice. She sucked on her bottom lip, and I could see she was thinking.

'Did you even find out her name yet?'

My toast popped up. I turned to butter it, slapped on some Vegemite, put on the kettle for some coffee.

'I don't know her name,' I said over my shoulder. 'I don't know anything about her.'

When I turned back around, a piece of toast sitting in my mouth, the girl was standing in the doorway, listening.

Her eyes looked wary, and she leant sideways against the doorframe.

'This is my daughter, Gem,' I said, biting off the piece of toast. 'Didn't get to introduce you last night.'

The girl's hair stuck out all bushy on one side and she held her plastered arm in her other hand. I wondered if she was still in pain.

'Rachel, that's my name.' Trembling, she spoke to the back of Gemma's head.

Gemma turned around and smiled.

'Hi Rachel,' she said. I could see she was uncomfortable. She put her bowl in the sink and pulled out some bread to make her lunch for school.

I popped a slice of bread in the toaster for the girl, and walked around the table to pick up the sling.

'Vincent,' I said, trying to figure out which way it would go on.

'Vincent,' she repeated blankly, still holding her arm.

'Come and put this on.' I motioned for her to come towards me.

She stepped forward, eyes like a frightened dog. The bruise on her face was going brown. I put the sling over her head and arranged it around the plaster.

'Vincent. That's my name,' I said, stepping back to the bench to butter her toast.

The girl stood in the middle of the room, her long wild hair pinned down beneath the sling. She didn't look feverish anymore, but sort of pale and small. I wondered if I needed to lead her to a chair.

Gemma stood beside me in the kitchen, making her sandwiches. She whispered, 'Dad, what's that weird smell? Like dirty socks or something.'

'Cabbage,' I replied, after a minute. 'Don't ask.' Turning, I called across to Rachel, 'What do you want on this toast?'

She looked at me but didn't answer.

'Vegemite?'

She nodded and moved to the table, awkwardly pulling out a rickety chair and sitting down.

'Coffee, or juice?'

She sort of opened her mouth, like she was going to say something, but nothing came out.

'Just give her some juice, Dad,' Gem said impatiently, getting out a glass. 'Do you want me to make you some sandwiches for work?'

'You got time? You need to shower and stuff?'

'This is early for me, remember. Bus doesn't come for ages.'

'Okay, sandwiches would be cool.'

I finished my toast, and poured hot water into the instant coffee in my cup. 'Hey, Gem, stay with her a second while I call up the doctor? I want to see if I can get Scott to open up early for me so I'm not too late for work. I better call the boss too.'

Gemma kept working on the sandwiches, but she nodded

her head. I grabbed my ciggies off the bench and took the phone out the back.

'So you've got our runaway, Vince.' Scott's voice on the line was muffled. I could hear his kids yelling in the background. 'The nurses told me about her when I did my rounds last night. They hoped she was alright. All she left behind were some spots of blood on the bedsheets. The coroner ordered an autopsy on the baby, find out the cause of death.'

I used to play footy in town with Scott as a kid. He went off to Sydney to do medicine, but he came back with his city wife to settle. They've got four kids now, a rowdy bunch.

'Get her name?' Scott asked, and then he yelled at his kids to quieten down.

'Rachel. She just told me this morning. Don't know her last name.' The air outside was crisp and cool. I lit a cigarette and sucked in smoke, filling my lungs.

'The cops couldn't identify her. There was nothing in the car, no ID. She didn't even have a purse. I guess they'll trace the registration, but I haven't heard.'

'She's pretty fucked up.' I peered around the doorway to make sure the girls couldn't hear me.

'Yeah. She's probably still in shock.' Scott paused. 'I'll drop in at the hospital to check out her file, then I'll meet you down at the surgery. Bout seven-thirty?'

'Great, thanks mate.' I hung up, then called work, said I'd make up the time in the afternoon. Stubbing out my smoke,

I walked back inside. Rachel sat chewing her toast, slowly, like it was cardboard. The smell of cabbage permeated the room. Gem snapped her lunchbox closed and headed for the shower. I heard the sound of the water draining from the bath and realised I'd forgotten to pull the plug.

'You want to have a shower after Gem?' I asked Rachel, wondering if she was bothered by the smell of the cabbage.

She shook her head, and took a sip of orange juice with a shaky hand.

'Maybe you should put on some fresh stuff then?' Opening the fridge, I crouched down to get the cabbage. I pulled off a couple more leaves and put them on the table beside her. 'I think you're supposed to keep changing them.'

She stopped eating and cradled her breasts with her good arm, a wounded-looking embrace. Her eyes were wide, as though the situation had suddenly dawned on her. As though she'd forgotten. Reaching her hand inside her shirt, she pulled out a cabbage leaf. It was limp and soft, as if it had been cooked. She put it on her plate next to her toast, staring at it while she pulled out the rest.

'It was all a dream,' she said. 'I was waiting to wake up.'

I could see she was going to cry again. Stepping up, I carefully pulled the hair out from under her sling.

'Aw, sweetheart,' I murmured, the words kind of slipping from my mouth. Smoothing her hair down, I felt helpless as the tears started dripping from her eyes. The milk began to seep onto her shirt, and she pressed against her breast with her open palm.

'Let's try and wrap them up again so you don't get soaked.' I got a couple of tissues and folded them over a few times. 'Put this in front first, then we'll put in some more cabbage.'

She lifted up her shirt and singlet, crying all the while, and I held the fabric and her plastered arm out of the way. Her breasts were swollen again. She struggled a little putting the cabbage in, but soon it was done. I hoped the tissues would be enough to stop her ending up drenched before we got into town.

'You ready?' I said, handing her a tissue for her eyes.

She held the white square between her fingers as though she'd never seen one before. Grabbing the sandwiches off the bench, I pulled on my boots at the door, patted my pocket to check for my cigarettes, and motioned her outside.

'Gem, we're heading off. See you this afternoon,' I called. Rachel didn't move. 'Come on, Rach,' I said, one foot out the door. 'You said you'd come to the doctor in the morning. You're not well, mate. You look real pale.'

Her glassy eyes were all accusation, as though she expected me to do her some sort of harm. 'The doctor. Not the hospital.'

'The doctor, I promise. He's an old friend. Scott. From the footy club.'

Pushing up from the table, she tottered towards me. I put my arm around her shoulder as we crossed the yard, and half lifted her into the truck. She winced with pain getting in, and I helped her click in the seatbelt. Throwing the sand-wiches in the back, ciggies in the console, I jumped in. All the way into town we bounced in the truck—the back shock-ies are shot—and I kept glancing at the side of her face to see how she was holding up. It was flat, her face, as though she'd pulled all the life in her deep inside, and every now and again she closed her eyes.

Wondering how the fuck I'd gotten into this situation, I reached down and pulled out a ciggie from the pack.

'You don't mind if I smoke, do ya?' I said, winding down the window.

She just shook her head and closed her eyes again. With one hand on the wheel I lit up my ciggie and turned to blow the smoke out the window. The bush sprang up green and lush, leaning in towards me. Through gaps in the trees I could see down across the valley, the cane fields spreading out before me in their neat endless squares. As we came down the mountain the curves of the road disappeared be-neath us, and soon we were on the edge of town.

Five more minutes, I thought, and then this problem would not be mine.

6

Scott met us outside the door of the surgery. Dressed in work clothes he looked different from when I drank a beer with him down the footy club. Slick, his thin wet hair neatly combed back, shirt tucked into cream slacks. He had a manila folder in his hands, and he tucked it under his arm as he unlocked the front door.

'Thanks, mate,' I said, pulling Rachel gently in after me.

It was strange being in the surgery early like that. No secretaries, no waiting. The heating hadn't been turned on and it was cold. The place felt kind of eerie. Scott headed straight for one of the rooms and I followed him in, hoping Rachel would just tag along. She wandered over but then hovered in the doorway of the room. Scott sat in his doctor's chair, peering down at the open file, flicking through the pages.

Sitting on one of the seats opposite Scott, I patted the chair beside me. 'Come on, Rach,' I said. 'Come sit down.'

Scott looked up at her then, and I could see him looking at the bruising on her face.

'So you found her on the road, you said, when she crashed the car.' He spoke to me, like she was a mute animal and he was a vet. 'And she hitched back to your house from the hospital?'

'Yeah.' I didn't like talking about her while she stood there, and I motioned for her to come into the room. She pressed her arm across her breasts and I could see the milk beginning to leak out across her shirt.

'You want to bring her inside, Vince?' Scott said, glancing back down and flicking over another page.

I stood up then and put my arm around her shoulder, pulling her into the room. She was trembling and I squeezed her tight for a second before I sat her down in the seat.

'So, broken humerus, head laceration, bruising. Does she have any painkillers?'

'I don't think so, she didn't have anything on her when she got to my house,' I replied, looking sideways at Rachel. 'I gave her some Panadol last night, though. It seemed to help. I think she had a fever. Her breasts are engorged.'

Scott looked up at me sharply, and then at Rachel.

'I rang up and asked my sister about it, cause they were leaking everywhere and she didn't want to go back to the hospital,' I said. 'Jan told me about the cabbage.'

'The cabbage?' Scott asked, tapping his pen against the open file and watching Rachel's face.

'You know, wrapping them up in cabbage leaves to stop the milk?'

Scott looked back at me and hid a quick smile. 'Yeah, right. The midwives always go on about that.'

He didn't look too convinced.

'Okay.' He pulled his chair over so he was sitting in front

of Rachel. 'She looks a bit dehydrated. Have you been giving her fluids?'

'I don't think she's drunk much.'

'Well, that's one of the reasons she was better off on the drip. Keep up the fluids.' He spoke to Rachel, 'Lift up your shirt there and I'll have a quick look.'

I turned away at the sight of her swollen breasts, still wrapped in cabbage, and stared at the posters on the wall.

'Yep, definitely engorged. Mastitis. I'll write her out a script for some antibiotics. But the main thing is she's got to keep expressing milk, just a tiny bit every few hours so blockages don't form, but not enough to stimulate the milk supply. It'll take a good few days for the milk to dry up. Probably need to buy some breast pads from the chemist.'

He looked at me and I wondered what I was supposed to say. 'Right,' was all I came up with.

Scott stood up to check the wound on her head. He pulled aside her hair, gently, and it flickered forward across her face.

'That's healing up well,' he said. 'I'll just take a quick look at her arm.' He moved the plastered arm around a bit, and then readjusted the sling. 'She's got to try and keep it in close to her body. Otherwise she's going to put too much strain on her shoulder.'

Rachel sat mutely, staring at something above Scott's head. Sitting down in his chair, he slid backwards across the floor away from her.

'Rachel?' he said in a loud voice. 'I need to talk to you

about your injuries, if that's okay?'

Rachel didn't reply, but she looked straight at Scott for the first time, suddenly still. I wanted to reach out my hand to her, but I didn't.

'Would you prefer if Vince leaves the room? I need to ask you some things that might be a bit difficult, but also a bit private.' Scott was watching Rachel's face very closely. Examining her. Then he looked at me, a questioning glance.

'I'll go.' I held up my hands. 'I mean, I'm late for work, I don't mind going.'

Rachel stood up then, real quick, and I knew she wouldn't stay if I left.

'Okay,' I said, putting my hand out and sitting back down. 'I can stay if you want.'

She sat down again, sort of perched on the edge of her chair.

'Rachel,' Scott spoke more softly now, 'when you got into hospital the other night the nurses in Emergency noticed that the bruising on your face looked a little older than it should have if you'd sustained it in the car accident, and that you had some bruises on your upper arms that resembled grab marks.' He didn't look away from her face. 'And when you went in for x-rays the doctor on call noticed that you'd had a few broken ribs in the past.'

I looked from Rachel to Scott and my insides lurched. I knew what it was Scott was trying to say. Looking at the bruises on Rachel's cheek, I flexed my hand on my thighs, hard until it hurt.

'Was the father of your baby—your husband, your partner—did he hit you?' Scott said finally. 'Because, though you don't have to tell us anything, if there is the smallest possibility of domestic abuse involved in the death of a child, we have to report it to police. That's the law.'

On the chair beside me, Rachel looked down at the floor. I could see her arm begin to shake, and I sat there watching her, knowing I should have seen it. She was so scared and I never even asked why. But seeing her face, I should have guessed.

'He never hit the baby,' she said after a minute. 'He only ever hit me.'

'Well ... the baby, what was his name?'

Rachel looked up at Scott, and I was surprised to see she wasn't crying.

'Frankie.'

'And how old was Frankie, Rachel?'

'Nine weeks. And four days, I think.'

'Well, the possibility of domestic violence will be considered when they carry out the autopsy on Frankie. I'm just telling you now because if there is any evidence that Frankie had been abused, it becomes a criminal investigation.'

Rachel gripped the side of her chair with her good hand.

'He never hurt Frankie,' she whispered, and Scott nodded his head, smoothing his fingers against his freshly shaved chin.

'Where'd you come from?'

'Sydney.'

'That's ten hours south of here. You drive straight?'

'I stopped when Frankie cried, to feed him and stuff. But I was scared to stop too long.'

I could see her breathing quicken.

'How'd you end up in front of Vince's?' Scott asked.

'I don't know … it was late. I was looking for somewhere to sleep, somewhere off the main road. I got lost and the road was so windy I couldn't pull off. It just kept going and going.'

'You had no ID on you. Was the car registered in your name?'

'No. It was his car. I left when he was asleep. I didn't take anything else, except the baby bag and some money for petrol.' She kept looking at Scott. 'I couldn't waste time getting stuff for me. He'd kill me if he woke up and found me packing.' Her voice cracked and she tucked her hair behind her ear.

'I think he must have taken my purse. I couldn't find it. He must have thought I'd try to get away. He didn't know I'd hidden some money.' She stopped talking for a second. 'It'll take a while for him to work out which way I've gone.' She looked over at me. 'Won't it?'

'The police will have traced the registration as part of their investigation,' Scott said. 'He may have already been informed of the accident.'

Rachel stood up, spine straight, black hair twitching down her back. I stood up too, not sure what she was going to do—she looked about ready for flight.

'I had to get out of hospital. When they worked out who I was I knew they'd tell him. I've been in hospital before, the time he broke my ribs—when I first got pregnant. I thought they wouldn't let him take me home, but they did.' Her whole body was shaking. 'Everyone always thinks he's nice, but if he finds me he'll kill me. He'll think I killed Frankie and he'll kill me.'

'Rachel, I'm going to call in Community Health. They'll hook you up with a social worker and get you into a shelter.' Scott reached for the phone.

'I'm not going to one of those places. He always said that's the first place he'd look. You don't know him, he'll find me.'

'It's a safe house, Rachel, it's set up for women in your situation. He won't find you there.' Scott's voice was firm and he lifted up the phone and dialled.

Rachel turned to me and put her hand on my arm, gripping it, her face close to mine.

'Vincent, he'll find me. I won't go. You can't make me go.'

I could see the sweat breaking out across the top of her lip. Her pupils were dilated, her eyes flicking about the room.

'Mate, what else you going to do?' I said softly. 'You've got no car, no money, you're hurt. Do you have family anywhere? Maybe the social worker can organise to help you get to your family?' I put my palms on the sides of her head, trying to get her to look at me properly. I felt like squeezing her head between my hands to stop her eyes darting around, but I didn't. I just pressed lightly and then patted her, trying to soothe her somehow, but she shook her head free.

'Rach?'

'I've got no-one.'

'No family anywhere?'

She shook her head. 'Vincent, he'll find me. He won't let me go.'

'Where were you headed?' I asked, thinking of the up-turned car on the road, lights pointing out to the trees, wheels spinning aimlessly in the air.

'I was just running,' Rachel said, and her wild eyes seemed to dull a little. She went limp then and I sat her back in her chair. 'Grabbed Frankie and ran,' she mumbled, and she looked like she might start rocking herself there on the chair. Like she was suddenly lost in some other place.

Scott hung up the phone and, watching Rachel, he spoke to me.

'Sue from Community Health will be round in a minute to pick her up. She'll get her into a shelter and help her sort out her prescription. I think they might even have a breast pump, so she won't have to do the expressing by hand.'

'Okay. Right. Sounds good,' I said, unsure how to respond. I kept looking across at Rachel, trembling on the chair.

'I reckon you're best off leaving while she's zoned out,' Scott said, rubbing the back of his neck. 'She seems to have decided you're a safe port. Might be hard to get out unless you go now.'

I smiled, though I didn't much feel like it.

'Will she be alright?'

'She's dissociated. It's the trauma.' Scott got up and moved towards the door, gesturing with his head for me to follow. 'A car crash is enough to knock anyone around. Post-traumatic stress, you know? It fucks up everything.'

We stood outside the door so she wouldn't hear, but she wasn't listening anyway.

'And then there's the violence.' Scott sighed. 'The body is a strange thing. I saw a woman the other day, assaulted once and she still couldn't taste anything six months later. Her hair stopped growing. Fingernails too. It's like everything goes into shutdown. And that's just the physiological side. People can get panic attacks and things for years afterwards.'

I watched her through the doorway.

'And she's copped it more than once,' I said, thinking about what that meant.

'Yeah, sounds like it. Domestic violence is a tricky area. But she's gotten herself free, that's the hardest part.'

'And now she's lost the baby.' I knew I didn't need to say it.

Scott nodded. He looked old suddenly, standing there in the empty surgery. Like a middle-aged man.

'The ladies from Community Health will take care of her, Vince.'

'Okay. I better go,' I said, wondering what to say to her. How to say goodbye. Walking back into the room, I crouched down and took her hand.

'Rach, I've got to go.' I gave her fingers a little squeeze. 'I've got to go to work now.'

She didn't look at me, just kept staring straight ahead.

'You look after yourself. Take care, okay?'

It all felt so useless, saying anything. The baby's slack face loomed up in my mind, and I dropped her hand. Frankie. His blue face and her milk dripping like tears.

'Bye, mate,' I said finally, but she just cradled her plastered arm and seemed to withdraw further inside.

I stood up, nodding to Scott. Then, grabbing for the smokes in my pocket, I softly padded out the door.

7

When I saw the woman all curled up in my dad's arms that morning, I knew there would be trouble. Standing in the doorway, watching them lying there together, I saw how she pressed herself against him in her sleep. She looked like a puppy nuzzling its mother, searching for something against the crook of his neck. She was different from his usual type, young and pretty, and wearing all her trouble on the outside. It shocks me how easily adults seem to fall into each other's beds. I was glad I'd slept through whatever went on in the night. There's nothing worse than listening to your dad have sex.

Dad denied it and he doesn't usually lie, but I kept thinking of the way she moved against him in her sleep and I just didn't know. Maybe Dad thought I'd tell one of the boys and then Marie would find out. Marie will think the worst anyway, she always does.

All day at school I thought about Dad and the broken woman. At lunchtime Dave came and sat with me. I was surprised cause he's never done that before. I could feel the skin of his leg against mine and I kept waiting for that strange prickling feeling, but it didn't happen. He said his mum was fucked and he put his arm around my shoulder, just for a minute. It was weird to hang out with Dave at school. When

I see him with Dad and Marie it's like we're part of some kind of strange family, but at school he's a boy and I'm a girl, and it feels different.

The girls I sit with at lunch moved away, and I could tell they thought something was going on. They were all watching me and giggling. Dave sat so close I could see that he had shaved. Tiny dots of hair were growing back above his lip, sparse and golden. My dad's real hairy, dark and kind of woggy. I wondered if Dave would ever grow hair like my dad.

Dave said even if Dad and Marie don't stay together we could keep hanging out, and I started to think that maybe he *likes* me. I've kissed boys a couple of times, but nothing real special.

I've never had a boyfriend, a proper one. Sitting there, I thought of how the woman moved against my dad, and I realised that I wanted Dave to kiss me. He didn't, but the rub of his leg next to mine as he shifted around felt meaningful. I wondered if I was going crazy.

Sometimes it feels like everything in the world echoes with a kind of meaning that's just out of my grasp. Things feel weighted and important, but then I wonder if it's all just me. Maybe no-one else notices. Dad doesn't seem to. It's the sort of thing that's hard to talk about. Sometimes I think that I could ask Dad about it, but I get stuck on the how. I could say, 'Dad, did you *feel* that?' or 'Dad, what does that *mean*?' But I think he'd just answer, 'What?' and then I wouldn't know what more to say.

When I walked up the driveway after school, she was there on the side of the road. She was crouching on the grass in front of the house, eyes closed, and when I got close I could hear that she was singing—the words slipping away just out of reach. I stood there and watched her, wondering what to do. I could have called Dad, but I figured I could handle it until he got home. She didn't seem to notice me standing there. She didn't even glance my way.

'Rachel?' I said, squatting down beside her. My school bag was heavy on my back, and I swung it off and onto the grass. 'You right?'

After a few seconds she stopped singing and looked across at me. It was as though she moved in slow motion. The bruises on the side of her face were like smudges of dirt.

'Frankie's dead,' she said, and she nestled her breasts in her good arm.

She was still dressed in Dad's trackies and old jumper, and she looked a bit mad, all stained down the front with milk. Her bare feet were dirty and roughed up, and her toes were bleeding. It looked like she'd grazed them on something. It's a long way to walk up to our place. I wondered if she'd been out all day, trudging along the road. Her hair was knotted up and wild at the back, and crouched down beside her I noticed for the first time how unusual it was—black and thick, and up close I could see strands of bright orange.

I reached out a hand and picked up a chunk of her hair.

'You've got red hairs in there,' I blurted out.

Rachel seemed to see me then, to really focus.

'Yep.' Her voice was quiet. 'About one in ten. My mum and I tried to estimate once. She was a redhead.'

'I've never seen that before.' I dropped her hair and stood up. 'You coming in?' She looked back at the road, and I held out my hand. 'Come on.'

Picking up my bag, I bent down towards her and she grabbed hold of my hand. I pulled her up, and with her plastered arm she scooped up a green bag that was beside her on the grass and followed me inside. She sat down at the table and I poured us both a cordial. I searched around in the cupboard for something to offer her, some chips or something. In the end I found some Pizza Shapes. Opening the packet, I tipped them into a bowl. She didn't touch her cordial, and I moved the glass closer to her. I felt like I should do something else for her, but I didn't know what. I was beginning to wish that Dad would come home. He'd have rung if he was going to be real late, so I guessed maybe he'd gone to see Marie after work. I didn't want to ring him there, because I knew Marie would freak out.

Rachel didn't move. I couldn't stand the silence so I started to talk. I told her about Dave at school and how the girls had giggled, and how I hadn't known if there was anything in it. I told her how I feel things but I never know if it's just me. I told her how I'm always grasping at the shape of things, but that just when I get it, everything changes. I told her about how sometimes the world seems to swell like waves around me like I'm pitching around in a storm. I told her about the feel of Dave's gaze on my back, how it made my belly surge.

The words seemed to spill from my mouth, and Rachel just watched me with her big dark eyes. Her eyes seemed almost bottomless, and I started thinking that she understood what I was saying. She didn't look away once. She didn't even seem to blink. I sipped my cordial and ate a few Shapes, pausing in my head to see if I'd got my thoughts out straight. She opened her mouth, like she was about to speak, but nothing came out. Suddenly the milk spread again across her shirt, and she pressed her arm along her chest.

I went to my bedroom and got my hairbrush. Standing behind her, I carefully worked the brush through the snags in her hair. The light was too dim for me to see the red strands, but I knew they were there. I brushed her long hair until it shone, the strange cabbage smell wafting up into my face. When I finished brushing I bent forward to look at her face. She was weeping again, noiselessly, her face still, like a mask.

I put the hairbrush down and I went to ring Dad.

8

After work I dropped in on Marie, thought I'd try to patch things up. She lives on the outskirts of town. White house, not much of a garden. The boys sat in the lounge room watching *Big Brother* or some other shit, and when I came in the front door Dave looked up and said, 'She's out the back, Vince.'

'Looks riveting,' I said as I went through. 'How can you watch that crap?'

The boys just shrugged. 'Shit all else to do.' Pete didn't even look up from the screen.

Out the back, Marie was working on some clothes. She'd cleared out her ex's old shed and now she used it as a sewing room. She did alterations and stuff for extra cash. Half the shed was hanging with clothes, and in the other half she'd set up her machine. She sat behind it, glasses on her head, a couple of pins in her mouth. The fluoro light was abrasive and I wondered how she stood it.

'What ya working on?' I said as I slipped under the roller door.

She looked up at me without smiling.

'One of the girls from Pete's year asked me to make her formal dress,' she said, pulling the pins from her mouth. 'It's a bloody nightmare. I've already unpicked this bit twice.'

I stood in front of the machine and peered down at the filmy pink fabric. It looked so delicate against Marie's hard, bony fingers.

'Must be a pain to sew, hey?'

'Yeah, slips around a lot.'

Pushing back her chair, she stood and held up the dress. It looked ludicrously pink in Marie's bright shed, and I couldn't imagine what kind of girl would wear it. Under the light I could see all the fine lines around Marie's eyes, and the strange mottled texture of her skin. She had silky hair, light-coloured and thick. The rest of her was a bit rough around the edges, but her hair always flopped about like it belonged to someone else. I loved it—it seemed like something precious—but she kept it pulled back in a ponytail mostly. I watched her, and wished she'd let it out.

Ignoring me, Marie smoothed out one of the seams on the dress, gently trying to pull it flat.

'Fuck, it's still fucking puckered.' She looked like she might start stamping her foot, a kid on the verge of a tantrum. 'What are you doing here anyway, Vinnie?'

'I missed you,' I said. 'You didn't have to leave last night.'

Marie's eyebrows drew in close.

'You can't expect me to hang around while you go all gaga over some young girl.'

I stepped towards her and took the dress from her hands, laying it carefully over a chair.

'I wasn't going gaga. She was fucking injured. You saw her, she was a mess.'

'She might have been a mess, but even Pete spotted right away that there was something between you two.'

I felt like grabbing her arms and shaking her.

'What the fuck are you talking about, Marie?' I was trying not to shout.

'Soon as she arrived Pete said something, I don't remember, something about it.'

I reached into my pocket and pulled out my smokes. Tapping one out, I lit it up and inhaled. I could feel my jaw clenching. 'Pete said she was pretty. That's got fuck all to do with me.'

When Marie carried on like this, it made me want to throttle her. Her husband left her a few years back, on Christmas Eve. They'd put a deposit on some presents for the boys, a dirt bike for Dave and something else for Pete, and she was waiting for him to come home so they could collect the presents, but he never showed. He cleared out their savings, took the car and pissed off with some other woman. The boys had a pretty shit Christmas that year. Their old man hadn't paid the phone bill, and the phone got cut off. Marie had to walk into town to call her mother, and some lady heard her crying in the phone booth and gave her fifty bucks there on the street so she could buy some food for Christmas lunch. I think she went down the pub and drowned her sorrows instead. Anyway, nowadays she's not big on trust. It drives me crazy. But there's something about her. I just want to be there. I put out my cigarette and then leant forward and gripped her shoulders.

'Marie, I feel like you got my heart down there under your heel.'

She wouldn't look at me, but stared down at her feet.

'Yeah, well, I got mine under my other heel, Vinnie.'

I ran my palms along her speckled arms. 'Well, why don't you fucking move, woman,' I whispered, bending down to kiss her. 'Lift up your feet and move.'

I pulled my fingers through her hair, pulling out the elastic and letting it down, and she let me kiss her. Then she pushed me away, and stepped back against the shed wall.

'You ready, Vinnie?' she said, looking at me, weighing me up. She lifted her denim skirt an inch and pulled down her undies, pushing them right down to the concrete floor. 'You ready now?'

She likes to do that, Marie does, like it's a test. She puts me off, and then pounces on me unexpected. And half the time I'm not ready, and that just fucks us both up. But I looked at the stretch of her legs underneath the fluoro light, backed up against the corrugated iron, and I felt myself go hard. Hallelujah, I thought, and I just dove right in.

Moving inside her, I could hear Marie's uneven breaths in my ear, and I closed my eyes and concentrated on the feeling of the thing, that building roar. And right when I was nearly there, inside my head I saw Rachel and her swollen breasts, and then I shuddered, done. I sagged against Marie, still gripping her thighs, and she pulled her legs free and slid her skirt down. When I looked up, she was watching me.

'You were thinking about her, weren't you?' she said, pushing against my chest and slipping past me.

'Marie.' I felt like all the blood was rushing to my head. 'You're fucking nuts.'

'I can tell when you're not really there.' She bent down to pick up her undies, all casual, like we hadn't just fucked, and inside my mind I imagined hitting her, hard. I pushed my hand through my hair, and pulled out another cigarette. Marie picked up the pink dress and suddenly my phone rang, loud and startling between us.

'Fuck,' I said, but I pulled it out of my back pocket. It was Gem's mobile. 'Gem?' I said. 'What's up?'

'Dad, she's back. The woman.'

I felt like punching that corrugated wall.

'Dad, she's been here a while. I tried to get her to eat something, but she's ... she won't talk or anything. She's crying.'

'Okay. Fuck it. I'll be there soon.' I hung up and lit my smoke.

Marie's eyes narrowed. 'She still there, Vinnie?'

'Community Health took her, but she's rocked up again. Gem's freaking out.'

Marie's whole body went stiff then.

'Get out of my shed.' Her voice was low and shaky. 'Get out.'

'Marie, come on. It's not like I want her there.' I moved

towards her.

'Nah, I don't want to hear it,' she yelled, heading out the door. 'Don't come round here looking for a fuck. I'm done with it.'

'Marie,' I groaned, watching her stomp away.

'Just fuck off,' she shouted back at me, slamming the door of the house behind her.

I stood in the shed a minute, dragging on my cigarette. Then I couldn't help it, I punched the wall, twice, hard. The pain spread out along my knuckles and up my wrist, and I shook out my fingers.

'Fuck it.' I stubbed out my smoke and headed out to the truck.

When I walked in the door Gemma was standing in the dark kitchen waiting, all big-eyed and sad. Rachel was crying soundlessly, perched on one of the chairs like a lost child. My hand throbbed, and I couldn't shake the memory of Marie's angry back. It seemed, just for a second, that it was all Rachel's fault. Her, and her dead baby too.

'What a fucking mess,' I said, my teeth clenching.

Lifting her arm to cover her face, Rachel flinched at the sound of my voice. It took me a second to realise she thought I was going to hit her. I didn't like seeing her that way, so I squatted there in the dimness beside her, reaching out a hand.

'Rach. It's Vincent.'

A scared kind of animal sound came from inside her.

She was grubby, all sticky down the front from the milk. I thought of the bruises and the broken ribs, and I felt bad then, real bad that I'd blamed her for anything.

'Rach. It's me. Vincent.' I tried to speak real soft. 'Come on, love, let me help you get cleaned up.'

She looked at me, and it was like my face suddenly swung into focus.

'Vincent?'

'Yeah, it's me.'

She glanced from me to Gemma and back.

'Come on, let's get you cleaned up.' I saw the green bag on the floor and peeked inside. 'Maybe you can try out the expresser thing?' There was medicine in the bag too. 'When'd you last take the antibiotic, Rach?'

She shrugged, one-armed, and I checked out the packet.

'Two doses gone. Think you'd be up for the third.'

Gem was hovering in the kitchen, quiet but jittery.

'Sweetheart, you want to grab a glass of water?' I called over to her. Popping the tablets in Rachel's mouth, I held the glass up to her lips. She swallowed and I could see her begin to shake.

'Drink up. Scott said you were dehydrated.' I was watching her eyes. 'Rach, you need to drink the whole thing.'

Looking at me, strangely steady, she swallowed and swallowed.

'When was the last time you ate?'

She didn't speak, just stared.

'Gem?' I glanced around at my daughter.

'Yeah, Dad?'

'Think you could heat up a can of soup? She needs to eat I reckon.'

'She walked all the way here. Her feet are pretty scratched up.' Gemma moved to switch on the lights. 'There's still a bit of glass out on the road too.'

I thought about what Scott had said in the surgery.

'She's in a state,' I said to my daughter. 'She'll be better when she gets some food into her.'

Gemma nodded, lips twitching like she was trying not to cry.

'It's okay, Gem, she's okay.'

I figured first things first—the girl needed a clean-up.

'Bath or shower, Rach?' I knew I might have to give her a hand either way, but I hoped maybe she could manage the shower by herself.

'All her hair fell out.' Rachel gripped my forearm, her fingers pressing into my skin. 'Just came out when you touched it.'

'What?' I was thinking about how to get her into the shower.

'My mum. All her hair fell out with the treatment. But I collected it up so it wouldn't be lost.' She pulled at the hair on my arm. I didn't know what she was talking about. It sounded like gibberish.

'Come on, bath or shower? I'll get a plastic bag.'

'Frankie's dead.' Her voice was low. 'He died too.'

I unclasped her fingers from my arm, but I let her words

soak in.

'I know.' I said it as gentle as I could. 'Frankie's dead. I'm sorry, Rach.'

What else could I say? I looked across at Gemma in the kitchen, wishing she didn't have to see so much pain.

'Now come on, Gem's going to heat you up some soup while I get you showered.'

Gripping my hand, she followed me into the bathroom. I pulled her shirt over her plaster, pants down her legs. I just wanted it all over quick.

Turning the shower on, I fixed the plastic bag in place.

'In you get.' My voice echoed against the tiles. 'Step in there.'

She stood shivering before me, her long hair hanging down. I wasn't sure what to do, what she was waiting for.

'Hang on, let me tie up that hair.' I figured it was no use having it all wet like the night before. I tried to sweep it all up in my hands, but it kept slipping out. 'Fuck, how do you girls do this?'

She didn't answer, and finally I got it tied and pushed her gently under the spray.

'Try to keep your arm out,' I said, looking away from her nakedness. 'And see if you can express some milk while you're there.'

I didn't know if she'd manage to get clean by herself, but it felt weird to linger in there waiting. I clicked the door shut and left her for a few minutes. Helped Gemma make some toast.

When I went back in, she was sitting on the floor of the shower capsule, humming, her plastered arm splayed to the side.

'Rach, what are you doing down there?' I was thinking of my grimy shower, the mould growing on the walls. 'Shower's not clean enough for that. Have to give it a scrubdown tomorrow.'

Reaching in, I turned off the taps. I'd grabbed some clothes of Gemma's, some old ones that she didn't really wear anymore, and a clean towel. Pulling Rachel up, I wrapped the towel around her.

'Gem's clothes.' I motioned to the bundle on the sink. 'She's taller than you, but I think they'll fit. No cabbage this time.'

She didn't move, just looked at me. Picking up the jumper, I put it over her head. It got stuck on her hunched-up hair and she started to squirm.

'Sorry, Rach, I'm no good at this either. Come on, we're nearly finished.' I bent down a little, holding the pants out so she could step into them. Grabbing my arm for balance, she breathed in like she was smelling my hair. 'Hey, you sniffing me?' I stepped back. 'I been at work all day. I must really stink.' Reaching for my ciggies, I glanced at the door.

'Cigarette?' Her voice was high.

Sometimes she said something like that, just out of the blue, and she seemed almost normal.

'You don't smoke, do ya?'

Rachel shook her head and water darted from her tied-up hair.

'Come on, let's get some food into you.'

In the end I had to spoon the soup into her like she was just a baby.

'Eat this, mate,' I said, tearing off a piece of bread and putting it into her fingers. 'Go on.'

She chewed the bread like it was plastic.

'I can't taste anything.'

'Just swallow it, Rach,' I handed her back the spoon, 'and then take another bite.'

She held the spoon stuck out from her closed left fist like a child, and I couldn't help but smile.

'You right-handed?'

Looking down at her arms, she paused a second then nodded.

'You'll get real good at using that left hand then,' I said, biting into my toast.

Rachel ate her soup, slow and clumsy, struggling to swallow. Gemma was watching me watching her and I could feel her mind ticking.

'How was Marie?' she asked, breaking the silence.

My hand still throbbed a little. I didn't want to think about it. 'Don't fucking ask.'

Gem looked down at her plate. 'What were the boys doing?'

'Watching TV. Slouching on the couch.'

Rachel put her spoon down and she seemed to sag a little to the side.

'Okay, Rach?' I asked. She hadn't eaten much, but maybe it was enough. 'Where's your sling?' I looked around the room. 'You need the sling on.'

She shrugged, shaking her head.

'Did she have it on when she came, Gem?'

'Nah, I didn't notice it.'

'We'll have to rig up a towel or something in the morning.' I finished up my soup. 'You want to lie down, mate? Leave the expressing thing till tomorrow.'

I went to get a doona to set her up on the couch. When I came back out from the bedroom she was standing up, waiting for me at the door looking scared.

'Come on, lie down. Gem and I will be right here.'

She shuffled towards the couch, watching me as though I might disappear. I put the doona over her and crouched down. She was dark under the eyes, tired-looking, but somehow alert. The bruises on her face were brown under the bare bulb.

'Close your eyes.' Standing up, I grabbed her discarded piece of toast, still hungry. 'Go on, close your eyes.'

Gemma finished the last spoonful of her soup. 'We should go to bed, hey? Turn out the lights. So she can sleep properly.'

I nodded, leaning against the table.

Gathering up the dishes, Gemma put them in the sink.

'Night, Dad.' She slid past me, closing her bedroom door. I turned out the lights, and stood there a while, until I heard Rachel's sleeping breaths.

Lying in bed, I kept replaying in my mind the scene with Marie. I just couldn't figure out how she could be so hard. In the end, I smoked a joint so I could get to sleep. But even with the pot, my thoughts raced and I shifted around, my old saggy bed creaking under me. After a bit I pulled out my stash and rolled up another.

I was finally sleeping when I heard my name in the darkness.

'Vincent.' It was soft, a whisper, and I struggled awake.

'Rach?' My voice came out slurred. I blinked, but I couldn't see anything. I felt her clamber onto the bed. 'You right, love?' I asked, still half asleep.

'Vincent?'

Maybe I shouldn't have, but I lifted the doona and she lay down alongside me. She burrowed close and I let her lie in my arms. I could feel myself edging back towards sleep, quick and tumbling. I don't know if I dreamed it, but I felt her put a hand on my chest, like there was something worth seeking under my skin. Like she'd captured my heartbeat under her hand.

'S'alright,' I murmured, my free arm lifting up to pat her head. 'I gotcha.'

And my arm dropped away in sleep.

9

I stood in the half-dark of morning and sucked in the smoke of my first ciggie, wondering if I should try and get Rachel out of my bed before Gemma woke up and saw her there. It was cold, nearly winter, and I felt my arms prickle with goose bumps beneath my shirt.

I didn't know what to do with her, this lone, battered girl. I thought about calling Community Health, but I knew it wouldn't do much good for them to come and get her if she just kept on running away. I didn't want to call the police or anything like that, but I knew I couldn't let her stay. I didn't even know her, and she needed help. Better help than I could offer. I barely had enough cash to keep me and Gem afloat, let alone another person. I only had five dollars in my pocket till payday, and I needed that for petrol to get to work.

I wandered towards the edge of the concrete path and bent to pick a few citronella leaves. Gem planted the little green cuttings one year to keep the mozzies away, and when you squeeze them they smelled real strong. I loved these plants, though they were sort of stringy and scraggly, untidy looking. Squeezing the leaves between my fingers, I held them up to my nose and then chucked them back into the garden, wiping my scented fingers on my pants.

Better than smelling like ciggies, I spose.

Stubbing out my smoke, I thought about how tightly Rachel had lain against me, gripping my arm in her sleep. How I had to unclasp her fingers before I could slip from the bed and creep out. I liked it, sleeping beside someone, wrapping someone up in my arms. I never really got to do it much. Marie and I didn't usually stay together. With the boys and Gem always around, it just never really seemed right. We fucked when we could, in those times when there was no-one about, but we didn't actually sleep together. I tried to block the thought of Marie out of my head, of how easily when we fought she always seemed to shed herself of me, but that just led me back to Rachel, and the feel of her clinging to me in the night.

Back inside, I made some sandwiches for lunch, spooning cereal into my mouth as I went. Sometimes I ate it dry, just to fill my belly so the coffee didn't go down too hard. The house was quiet and I tried not to make much noise, but I soon heard the sound of Rachel stumbling out of bed and then bumping against the doorway.

'Shh.' I held my finger up to my lips as she appeared in the kitchen. 'Don't wake Gem.'

Rachel nodded, rubbing her eyes with her good hand.

'I got to go to work, Rach,' I whispered as I switched on the kettle. 'I start at seven, and it's kind of out of town.'

She cupped her breasts, and stared at me with her large eyes.

'They sore?' I asked, pointing at her chest.

She pressed her arm against herself and nodded again. The bruise on her face was just a shadow.

'You want to try out the expressing thingamy?'

She walked over to the couch and sat down, and I handed her the green bag. She sat there staring at the wall, blank-faced, while I made her some toast.

'You could come to work with me if you want. It's green and peaceful out there. A citrus orchard. I do the maintenance with a couple of young blokes.' I said it without thinking. Soon as I spoke I realised the ribbing I'd get at work for bringing along a woman, specially one like Rachel—banged up and pretty, with leaking breasts and all. I'd never live it down.

She nodded and I made myself a coffee.

'Or you could just hang out here?' I asked, a bit hopefully.

She shook her head, cradling her plaster, holding the weight.

'You sure? Might get a bit boring out there.' Handing her a piece of toast, I started collecting up my things. I knew she wouldn't stay at home without me.

'You want to try the expressing thing before we go?' I asked again, thinking of her whipping the breast pump out in front of the fellas. 'I'll try and find a towel we can use as a sling. Guess it needs to be kind of square, hey?'

Pulling the plastic machine out of the bag on her lap, I handed it to her, taking her toast and putting it on the table. Then I went to the bathroom to look for something to use for a sling. Riffling around in the cupboard, I found a squarish thin bathmat that I thought might do the trick.

When I came back in she was sitting on the edge of the couch, jumper tucked up, the funnel end of the breast pump over her breast. I watched her a second from the doorway to see how she was managing. The plastic contraption had a tube that led to a small balloon-shaped pump, like a blood-pressure machine. Rachel had to squeeze the pump with her good hand and hold the funnel end in place with her bunged-up arm. I restrained myself from kneeling down and taking over. There was only ten minutes or so before we had to go, and I caught myself thinking maybe she could do the expressing in the car. After a few minutes she seemed to get it going and I could see her breast stretching out inside the clear funnel. The suction looked painful and I had to turn away.

'Let's not forget the antibiotics, hey?' I said and fished them out of the green bag. I went and got her a glass of juice.

'Vincent, it's not working.' Her morning voice was all husky.

I crouched down in front of her. 'Why not?' I said, checking the tube was connected right.

'I can't get a let-down.'

'A what?'

'When the baby sucks, it sets off a let-down and the milk just flows. You feel it like a sensation, like a dam breaking or something, it's really strong. The milk won't come out without it.' She looked like she might cry.

'But your milk has been constantly leaking out,' I said, and realised how blunt it sounded.

'I know, but it was like Frankie was just here in the room, I could almost feel him against me.' She pulled the machine off her breast. 'Now it feels like he's gone.'

Her breast sat swollen and sad-seeming, right at my eye level. I could feel the weight of her grief damming up in the room.

'Thinking about your baby is supposed to give you a let-down, but thinking about Frankie now just makes him seem more far away.' Her voice cracked.

'But you've still got milk?' I asked, standing up and looking away.

'Yep, they feel full.'

I didn't know how to help her, and I stood there with the glass of orange juice in my hand and thought about the minutes ticking by. If we didn't leave now I'd be late for work.

'Let-downs are funny sometimes, Vincent.' Her voice was low. 'They can be set off by the oddest things. One day I was buying milk from the corner store, and I put the carton up on the counter and the man said, "Just the milk, love?" and I got a let-down right there in the store. Two big wet patches on my shirt. I laughed, but the man didn't think it was funny.' She smiled then, and I couldn't help but smile back.

'So you got a let-down from a carton of milk?' I asked, and suddenly I had an idea. Stepping across the room, I tipped the juice I was holding down the sink, rinsed out the glass and filled it up with milk. I grabbed another glass and stood in front of her. 'Okay, so rig it back up and try again,' I said, and she put the plastic funnel back over her breast.

I held the glass of milk in front of her, and slowly tipped it sideways, pouring it into the other glass.

'Start pumping. The *milk* is flowing from one glass to the other.' I poured it back into the first glass. 'The *milk* is flowing. The *milk*. The *milk* is flowing from one glass to another,' I said in my best hypnotist's voice.

Rachel smiled again, her gap-toothed grin, and then like magic I saw the milk begin to flow. It seeped down into the plastic bottle, noiselessly, quickly, as though it would have even if she wasn't pumping. Then I saw the other breast was leaking too, spraying out like a fountain. Rachel's face seemed to change, like a dark veil dropped over it, and she breathed out in a heavy sigh.

'A let-down,' I murmured, putting the empty glass on the table and pulling out a few tissues to put over the breast that was spraying. I shook my head. Life seemed stranger every day.

While her milk filled the plastic machine, I grabbed the sandwiches I'd fixed up, made an extra couple for Rachel, found us both a plastic cup, and then tried to figure out how to work the sling. It had to be folded into a triangle, I knew that much.

'Rach, we really got to go. I'm late.'

She lifted her face up to me, but her eyes were sort of cloudy. Pulling the breast pump off, she held the tissues absently against her spilling chest.

'Here, take your tablets and drink this milk up.' I handed her the glass and popped the antibiotic into her mouth. She drank the milk, all in one go, and handed the glass straight back to me.

'Let's try the sling.' I pulled her jumper down over her breasts, glancing away from her dark nipples. Folding the triangle of the towel around her plastered arm, I tied the ends together behind her neck. The sling seemed alright, a bit high maybe, but Scott had said it had to be tight.

'Okay. Grab your toast and let's go.' I picked up the pillow and doona from the couch and headed out the door. Rachel followed me, holding her square of cold toast limply in her free hand. Throwing the doona and pillow in the back of the truck, I ran back inside for the cups and sandwiches. I spied Gemma's old sneakers and picked them up on the off-chance they might fit Rachel. It was too cold for bare feet. At the last minute I shoved the antibiotics in my pocket, next to my ciggies, realising we wouldn't get home in time for her next dose.

Outside, Rachel was already in the truck, holding the toast in her mouth and struggling with the seatbelt. I jumped into the driver's seat, leant over and clicked her in, and then started the engine.

The boys didn't say anything when I arrived late and set Rachel up on the doona in the sun. The air was crisp out there this time of year, specially if you weren't moving around. I thought she might get chilled. I could see Steve and Jimmy glancing between themselves, but they just kept on working, trying not to watch me fussing with the blanket. The boys were clearing a bank that was overgrown with lantana. A small fire smouldered, and they were dragging woody clumps of sticks from the bank straight onto the fire. Lantana's like that, so dry you can easily burn it as you go. Rachel sat down on the doona and I started straight in pulling out the lantana with the others. I kept my eye on her as I worked, and in a while she lay down and fell asleep. We stopped working for a sec to have a smoke. That's when the ribbing started up.

'Vince, where'd you find a girl like that?' Steve asked, punching me on the arm. He fancied himself a bit of a ladies' man.

'She's too hot for you, mate,' Jimmy added with a grin. 'Did you knock her around a bit or what?'

'Fuck off,' I said, lighting up my ciggie. 'I don't fucking hit women.'

'What's wrong with her anyway?'

'She's had a hard time. She's just a friend. I'm looking after her for a few days.'

'Gee, I bet Marie's chuffed,' Jimmy said, wiping the sweat off his forehead with his shirt. He's a big bloke, Jimmy, but he's not cocky like most young fellas. He watches stuff.

Doesn't miss much.

'Last time I saw you two out on the town Marie was going nuts about some woman and you were trying to calm her down. This her then, Vinnie?'

'What?'

'This your other woman? The one Marie's always psyching out about?'

I sucked on my ciggie hard and then looked across at Jimmy.

'I don't have no other woman. Just Marie. Rachel's a friend. Anyway, she's probably not much older than Gem. I don't sleep with young girls, Jimmy. Unlike you cowgirls, I'm nearly forty.'

Jimmy tried not to smirk. 'Alright, mate, whatever you say.'

'Now that you mention it, Vince,' Steve broke in, 'she does look more like my kind of girl. Where's she from? Looks like some kind of mix. Is she part Asian or what? I like those sweet little Asian girls. Indian maybe?'

'She looks a bit like one of them island girls, like from *Mutiny on the Bounty.*' Jimmy added his two bobs' worth. 'Where was that movie set again? I saw it on telly the other day, must have been on cable.'

I just shrugged. I'd never seen the movie, and I didn't know where Rachel was from. She talked like an Aussie.

'Introduce us later, hey? Maybe she just needs to have some fun, you know, with someone her own age.' Steve ground his pelvis in the air as he spoke. I knew he was just trying to wind me up, but it was pissing me off.

'Just leave her, right, boys? She's way too much trouble for either of you to handle.' I chucked my smoke on the fire. 'Let's finish this bank, then we can cut off some of those dead branches,' I said. 'We don't want the old guy to rock up this arvy and see you've spent all day gawking at a sleeping girl.'

At lunchtime I woke her, thought she might need a drink and her next dose of antibiotics. She sat up awkwardly, one-armed and off balance.

'You right, mate?' I asked, but she just looked at me. 'Come and have something to eat—if you sleep all day you won't sleep tonight. Here, try these shoes.'

I helped her put Gemma's shoes on, and they seemed to fit okay. Stumbling up, she grabbed my hand and I brought her over to where the boys were eating lunch. They sat slumped on discarded logs, munching from a big bag of chips and drinking Coke. Rachel's towel-sling looked bulky and uncomfortable, and I wondered what else I could have found for her.

'Rachel, this is Steve and Jimmy,' I said, hoping they would behave. 'Boys, this is Rach.'

Jimmy held out his hand for her to shake, and she stared at it a minute before letting go of my hand. She only had her left hand free, so Jimmy swapped arms and they clumsily shook hands. When Jimmy touched her he went quiet, and I knew then that he saw she was not quite right.

'So,' Steve said brightly, 'where you from then, Rach?'

Rachel looked across at me before answering. 'Sydney.'

I sat down and pulled her gently onto the log beside me. Unwrapping a sandwich, I passed it into her good hand.

'And what brings you all the way up here?' Steve kept up, tipping chips into his mouth.

Glancing from Steve to Jimmy, Rachel's words came out swallowed and soft.

'I was running.' She smiled, a wobbly trembling smile, and the sandwich sagged in her fingers. 'My mum was sick, and when she died, well, there wasn't anyone except Paul.'

Both Steve and Jimmy looked at me, a bit panicked, and I shrugged. It was all news to me.

'All her hair came out. It was red, see? I left it behind when I ran, but I thought I'd have Frankie to remind me.' Her voice sounded wet, on the verge of tears, and the boys' eyes darted away. 'When he was born, no-one thought he was mine, his hair was so red.'

'Hey, Jimmy,' I picked up the empty plastic cups, 'can you get us some water from the tap, mate?'

Jimmy jumped up with relief.

'Eat your sandwich, Rach. Then I can give you your tablets. I don't think you're supposed to have antibiotics on an empty stomach.'

Rachel took a bite and slowly chewed.

'I need to do a wee,' she said. 'Is there a toilet?'

'Nah, just pop over there behind that bush.' I pointed a

little way off. 'No-one will see, I promise.'

She handed back the sandwich and wandered towards the bush.

'Is she alright, Vince?' Steve said when she was out of ear-shot. 'She doesn't sound too good.' He was a youngish bloke, Steve, but all of a sudden he seemed just a kid.

'She's been through some stuff.' I bit into the sandwich. 'She'll be right soon.' But even as I said it I stood up to watch if she was coming back. 'She's just a bit lost, that's all.'

Jimmy stalked over from the tap with the cups full of water. 'I don't know where you find them, Vince. I thought Marie was a handful,' he said, handing me the cups.

I didn't say anything, just stood there keeping an eye out for Rachel.

'She sounds a bit nutso. Is she a bit crazy?'

I felt the anger inch up my back.

'She's not fucking crazy.' My hands clenched around the plastic mugs, squeezing them out of shape. The boys went silent, their heads down. 'She's not crazy, alright? Just hurting real bad.'

'Okay, Vince, okay.'

Jimmy lit up another cigarette and took a swig on his Coke. 'How'd you meet her then?'

I watched Rachel come out from behind the bush and walk towards me, her dark hair wisping around her in the wind.

'It's a long story, boys. A long fucking story.'

She was restless, wandering round the orchard touching the fruit, on the verge of some big emotion. It was making the boys nervous, so I thought it'd be better if I got her moving.

'You want to help too, Rach? We're going to do a bit of pruning. You can carry some sticks to the fire for me.' She looked at me, and I pointed towards a branch. 'Some of these trees have got gall wasp. See here, this little bulby thing. That's the gall. We got to cut the branch off if it's got those lumps cause it means the gall wasp has got in and the branch will die.'

I picked up the secateurs that were lying in the grass. 'I'm going to cut them off and hand them to you, and then you go and pop them on the fire over there. Then they'll all be burnt off before we go.'

Grabbing the branch with the lump, I snipped it off and handed it to her. She stood staring at it for a few seconds and then slowly walked over to the fire and placed it on top. I kept working, piling the sticks up on the ground for her. One of the boys started the chainsaw over the other side of the paddock and I saw Rachel flinch.

'It's just the boys with the chainsaw,' I called out to her. 'There's some bigger branches down that way that have to go.'

The noise was loud and jarring, and she looked a bit stunned.

'Come on, Rach—me and you, we're sticking to the small stuff.'

I held out a stick to her and she moved across to get it.

'Gall wasp,' I heard her mumble under her breath as she

walked back towards the fire.

Chopping those branches, I kept my eye on her. She was standing by the fire, singing softly. I could hear fragments of her voice, but I couldn't catch the tune. After a little while she stopped, put her stick on the coals, and headed across to grab some more. I passed her on the way with a big bundle of branches.

'Whose place is this, Vincent?' she asked, looking around.

Some of the autumn leaves were still on the trees up the back, and I thought she might like to see them.

'Come on, I'll show you something,' I said, grasping her hand so she'd follow me. The afternoon light shone through the coloured leaves. Even I could see it was beautiful, but the sight seemed to hurt Rachel's eyes. She looked away, and I saw how much the sling was cutting into her neck.

'The towel's too heavy, isn't it?' Reaching over, I lifted her hair, adjusting the knotted towel behind her neck. I'd done a shit job. I could see it hurt, but there wasn't much I could do.

'The old guy who owns this place,' I said, bending to pick up a yellow leaf from the ground, 'he loves these trees. Liquidambars they're called, and even though they're killing the fruit trees underneath, he won't cut them down.'

The fruit trees around us were stringy, sagging with heavy fruit. Weeds grew round the hunched trunks, fallen yellow leaves caught between the branches.

'See, they look like ghosts, struggling here in the shade.' I kept on talking. 'He makes us leave them, no fertiliser or nothing, and look, they're covered in mosses. That one's even got a staghorn growing on the bottom. Epiphytes and all.'

Reaching out, I pulled off an orange fruit. 'The old guy loves these scraggly trees even more than the liquidambars. Won't let us cut them down either. I reckon the whole place is some kind of tax write-off anyway. Hobby farm my arse. Only some rich old city bloke would be able to afford to pay three men to do all this shit. But it could be worse—I could be still packing boxes at the hardware. At least it's outside.'

I threw the fruit up and caught it.

'This one's a mandarin. You want one?' I figured it would be good for her. Bit of fresh juice. 'I'll peel it for you.'

Pulling the skin away, I tore the mandarin into two halves. Inching off a piece, I bit into the pith, pulling out the seeds with my teeth. Just like I used to do for Gem when she was little. 'Here you go. No seeds.'

Rachel opened her mouth like a bird. As she bit down, her eyes filled up with tears.

'Bit sour, hey?' I had to smile. 'Not quite ripe yet.'

I threw the fruit over her head and into the trees, and headed back towards the secateurs. She trailed behind me, and from the corner of my eye I saw her catch a yellow leaf as it fluttered down.

'Just a few more trees, love, then we'll find another job.'

Dropping the bright leaf, Rachel wandered back to the fire. She looked peaceful standing there, so I crossed over to work on some trees a bit further away.

I didn't see her take off her shoes, and she didn't make a sound when she stepped in. I just glanced around and saw the lines of smoke floating up on the wrong side of her body. It took a second for me to realise what she'd done, and then I was running.

We hit the ground hard, the jolting thwack smacking in my ears.

'Fucking hell, Rach!' My voice felt strangled and raw.

I scrambled up, holding her, and sprinted across the orchard. Dropping her on the ground beside the tap, I turned the water on, pulling her feet beneath the spray.

'Fuck.' The water was cold, and it spread upward along the cloth of her pants, splattering up at her face. 'What were you doing? Trying to fucking kill yourself or what?' I was yelling. The sound of the rushing water was loud. 'You're not going to die stepping into a fucking fire, Rach, you're just going to get burnt.'

She held up her good arm to shield her face, and I saw how angry I must have seemed.

'I'm not going to fucking hit you, okay?'

My fingers shook as I pulled the material of her pants up her legs and out of the streaming water. I lifted her foot up to look at her sole.

'Does it hurt? It's a bit red, but not too bad yet.'

Rachel shook her head. Somewhere deep down I knew that there were other types of pain.

'What did you do that for?' I asked softly.

Her face was full of loss, her lips trembling at the corners.

'I just wanted to get clean.' Her eyes spilled over. 'Vincent, it's my fault, isn't it? That Frankie's dead. It's my fault.'

I didn't know what to say to that.

'It's a bad corner, Rach.' That was all I could come up with. I knew it wasn't much, so I just let the water run.

'I've got to get you to hospital, I reckon,' I said finally, turning the tap off and lifting her up.

'Not hospital. The doctor.' She cried then, thick gulping sobs.

'We'll see, Rach, we'll see.'

Carrying her across to the truck, I lifted her in, then ran back to the tap and wet my shirt so I could wrap her feet in it. Her soles were already blistered. Running around to my side, I jumped in. She was sucking in wet breaths, hard and fast.

'I'll ring Scott on the mobile, but if he says hospital that's where we have to go.' Searching for the number on my phone, I veered a little on the road and I slowed down to make the call.

'Scott, it's Vincent.' I had to shout over the revving sound of the truck. 'Listen, are you at the surgery?'

Rachel's breath seemed to catch in her throat. She lifted her hand to her mouth.

'Okay. Well, I got Rachel here ... she's burnt her feet ... stepped on some coals ... blistered. Yep. To Emergency? It's just, she's freaking right out. About the hospital.'

I was listening to Scott's reply, but I could see her starting to struggle.

'Shit. I think she's hyperventilating. Hold on.' Dropping the phone, I pulled off the road. 'Rach, calm down. Fuck it. *Calm down.*'

A high-pitched squeaking noise came from her throat, and she grabbed at her jaw. I bolted round the back of the truck, looking for something to give her. Pulling out an old bakery bag, I blew into it and then held it over her mouth.

'Okay, now breathe slowly. Breathe out.' I bent down to pick up my phone from the floor. 'Scott?'

Rachel was still sucking in breaths, and Scott's voice seemed very far away.

'Yeah, she's right. Emergency? Okay. Thanks, mate.' Standing shirtless in the open car door, I held the bag for her until her breathing slowed.

'Scott says hospital. He's playing golf,' I said quietly, watching her. 'Okay? Now you hold the bag, while I get us into town.'

Pulling back onto the road, I fumbled in my pockets for my cigarettes, but I'd left them behind. I stepped down on the accelerator and the truck surged forward. Rachel sat there breathing into the paper bag, limp and heavy-eyed.

10

As I drove up the hill towards the hospital I could hear Rachel's breathing quicken. I couldn't help but imagine her running down the same hill a few days back, bare feet and bleeding, trying to get away from where I was now taking her. It made me think about what her boyfriend must have done to her to make her so scared. The blood banged in my ears, and I gripped the steering wheel.

'Rach,' I said, as gently as I could. 'It's alright, baby.'

Unclenching my fingers, I reached out and put my palm on her shoulder. She was shaking, hard.

'I don't know how bad those feet of yours are, but we have to go and find out. You know that, right?' I asked, though I didn't expect an answer.

I pulled up outside Emergency, gathered her up, and pushed the door open with my shoulder. Carrying her through the waiting room, I found the nurses' station but couldn't see anyone around.

'Got a girl here, she's been burnt,' I called out.

Rachel's breathing sounded ragged, and she still clutched the paper bag to her mouth. I heard the click of the nurse's shoes, hard on the cold floor.

'Vinnie, who you got there?' It was Norah, an old friend

of Marie's. I'd forgotten she worked at the hospital. I couldn't believe my luck. It's not like I know any of the other nurses. Fucking small towns.

'Her feet are burnt. She stepped into a fire,' I said, not wanting to say more.

'You'd better bring her straight in.' Norah pointed through the door to the Emergency ward. Inside, the beds were all empty, and I laid Rachel down on the closest one.

Norah stepped over to the phone on the wall and murmured quietly into the receiver. Coming across to the bed, she bent over Rachel, unwrapping my wet shirt from her feet and examining her soles.

'Doctor will be round soon.' Norah glanced up at me while she spoke. 'I'll clean them up in a sec. Burns on the feet are really painful. I'd say she's probably in shock.'

I looked at Rachel lying on the bed. She was wet and covered in goose bumps, with sooty patches of black smeared on her face. The towel-sling had come half undone, and one of the corners hung down against her belly. She stared vacantly to the side, quiet and pale, her whole body convulsing slightly every few seconds. I picked up her left hand, it was limp and clammy. Giving it a squeeze, I bent down to say her name, but she didn't respond. She didn't even turn her head.

Standing there, I could feel Norah watching me, taking in my bare chest and wet clothes. Watching me hold Rachel's hand. I could feel her judgement moving through the air and landing hard on my shoulders, and I wondered if I was covered in black smudges too, marked out for my dark deeds.

Norah chucked my shirt in the bin and covered Rachel with extra blankets, tucking them in around her.

'I'm just going to take her obs. Before the doctor gets here,' she said, picking up a clipboard attached to the end of the bed. Reaching out, she took Rachel's hand from mine, feeling for her pulse.

'What's your name, love?' she asked Rachel, peering down at her face. If Rachel heard, she made no sign. 'What's her name, Vinnie?'

'Rachel.' I looked down at her on the bed, wanting to fix her up a bit but not knowing where to start. I brushed her hair back from her face. Tried to wipe off one of the black marks. She stared blankly, not meeting my eyes. 'I don't know her surname. She was here last weekend.'

Norah looked at Rachel sharply. 'The girl who lost the baby. That's what the arm's from—the car crash?'

'Yeah.'

'Everyone's been wondering what happened to her.' I could feel Norah's gaze on my face. 'She been up at your place, Vinnie?'

I didn't really want to answer. I knew whatever I said would go right back to Marie.

'I'll jot you in as the emergency contact, then?' Norah looked down at the clipboard, holding her pencil.

'Sure,' I said, trying to keep my voice normal. 'Why not?'

I knew what she was thinking. Same as Marie, same as

Gem. I felt the blood surge up into my head. I wanted to rip the clipboard from Norah's hands and snap it in half, but I just stood there in the dingy hospital room and watched her write out my name and phone number in her neat nurse's writing.

After she'd fussed around a bit filling in Rachel's chart, Norah gathered up some implements—dressings and bandages, plastic tweezers and some sort of salve—and walked down to the end of the bed. Lifting the blankets she looked at Rachel's burnt feet again.

'You know, Vince, they don't look so bad. You obviously got her under some water, hey?' she said, reaching out and touching them with her finger. 'They're blistered, but you'd expect that.'

'Yeah. I guess.'

'She's definitely showing signs of shock. Blanked right out. Blood pressure's a bit high. Sometimes surface burns are more painful because the nerves are still intact.' Norah's voice was casual, just another day at work. 'I reckon the doctor will want to admit her, at least for a couple of days. Sometimes they let them stay a little longer if it's not busy and it'd do them good.'

Rachel didn't even blink. It was as though she couldn't hear.

'What were you doing?' Norah asked me. 'How'd it happen?' She picked up Rachel's feet and put them on a blue plastic mat. Snapping on a pair of tight gloves, she broke open a bottle of saline solution, squirting it on the blistered skin.

'I don't know. We were burning off some stuff. I turned away, then turned back and she'd stepped in the fire.'

'She stepped in on purpose?' Norah stopped squirting the saline and looked at me.

'Said she wanted to get clean.' I felt suddenly as though I was covered in dirt. My eyes filled up. Clenching my jaw, I covered my face with my forearm, just for a sec.

'She's scared of the hospital,' I said, pulling my arm away. 'She was on the run, see, when she crashed the car. When the baby died. She's scared her bloke will come and find her.' My voice came out soft and husky, but I held it together alright.

Norah shook her head, breaking open another container of the sterile water and spraying it on with her gloved hands. 'Fucking men, hey?'

I could only agree. 'Yeah, bunch of bloody bastards.'

Norah laughed, a sad sort of laugh, but I couldn't even smile.

The doctor came in, starched plaid shirt and smelling of disinfectant. He was straight out of uni, I reckon, one of those young city blokes that come out this way to get their hospital hours up and then piss off quickly. He introduced himself and bent down to look at Rachel's feet.

'Yeah, just put on some Silvazine, a dressing and bandage,' he told Norah. 'The burns aren't too severe but she looks a bit the worse for wear. She was in here last weekend, wasn't she, car crash? Baby DOA?'

'Yeah,' I replied.

'Scott called me, told me she was on the way in. I checked out her file. Mastitis too. Post accident.'

I felt around in my pants pocket and pulled out the antibiotics, handing them to the doctor.

'Has she been taking them?'

'Yeah, probably not real regular.'

'She looks a bit dehydrated. Considering the trauma of the last week I think we should keep her in here for a few days. Just get her back on her feet, so to speak.'

I watched Rachel as the doctor spoke, waiting for her to freak, but she just lay there staring at the far wall.

'Put in a line for fluids, then five milligrams of morphine for the pain,' the doctor said to Norah. 'Then we'll hook her up some IV antibiotics.'

Norah nodded and moved away to gather the equipment.

'Look, can I have a word?' I said to the doctor, motioning outside with my head.

The doctor looked at me a second, weighing me up. 'Sure,' he said, and walked towards the door.

'You've read her file?' I asked when we were out in the hallway, not quite sure where to start. 'You know the bruises and stuff weren't from the crash? Her bloke had been banging her around.'

He breathed out, a tightness leaving his body.

'Yes, I am aware that she is in all likelihood a victim of

domestic violence,' he said in his slick city tone. 'I assume you're not her partner then.'

I realised that he'd thought it was me, he'd thought that I was the one who'd been hurting her.

'Nah, didn't Scott tell you?' I clenched my fists again, it was all too hard to explain. 'I'm just a friend.'

'Okay, Vincent. That's what you said your name was, right?'

'Yeah. Look, I been struggling to get her to come back to the hospital, and she just flips right out cause she's sure her boyfriend is going to come find her here, to come get her. I just want to know, is she going to be safe? Is there some way she'll be protected? If he does come?'

'The police have been looking for her, to interview her again about the accident. They tried when she first present-ed, but she was too muddled. Then she ran off and they hav-en't been able to track her down.'

'She was up at mine. The cops took photos of the road up there, they know what it's like.'

'It's just procedure. Casualty from a single MVA. They have to follow it up for the coroner,' the doctor said. 'If she needs protection, I assume they'll set it up.'

I knew a couple of the cops in town. Unless it was an emer-gency, they weren't in any hurry. Bunch of bloody wombats. They'd take their time.

'I guess in the meantime I could put a note at reception that she isn't to receive any visitors,' the doctor continued, glancing at his watch. I wondered where he needed to be.

'You reckon that will be enough?' I asked. I'd been to the hospital to visit friends before. I didn't think it would be tricky to slip in unnoticed.

'Look, I'll alert staff to the situation, but really it's a police matter.'

'Right.' I was unsure what else to say.

'Listen, Vincent, this is just a small country hospital. I think it's unlikely that her boyfriend would get past the nurses without their knowledge.' I could tell he wanted to get moving to wherever he was going next. 'She's obviously suffering some level of psychological trauma. She'll have access to a social worker here who will hopefully be able to help her.'

'If you can keep her here,' I replied, looking at the doorway. 'She's done a runner twice now. Straight up to my place.'

'Well, I'm afraid we can only offer her our services, we can't force her to accept them.' The doctor ran his hand along the top of his neatly cropped hair. 'The burns on her feet are going to keep her here for a few days at least. Maybe that will be long enough for her to reconsider her options.'

I looked at the young bloke's face and wondered about Rachel's options.

'I've got to keep doing my rounds on the ward,' he said. 'Norah will fix her up, and I'll check in on her again on my way out.'

The doctor walked away and I went back into the

emergency ward. Norah was finishing the second bandage, sticking it tight with tape. She looked up at me as I walked back in.

'I'm just going to get her out of these wet clothes and into a gown, if you just want to wait outside a sec.'

Rachel whimpered then.

'No.' Her voice was hoarse. 'No, don't go again, Vincent.'

'Norah's going to get you changed,' I said to her, watching her face, not stepping any closer. Her eyes looked a little wild. I could see she felt trapped.

'Stay. You help me.'

Norah shrugged, turning away. 'I'll just get the IV ready then,' she said, handing me a crisp white hospital gown.

I stepped up towards the bed and Rachel started wrestling with her jumper.

'Wait. Let's get the sling off first.' I reached around beneath her hair and untied the knot. The towel came away in my hands and I chucked it in the bin on top of my wet shirt. I pulled Rachel's sleeve over her plaster and then carefully over her head. Her skin looked pale under the fluorescent lights, and I wished I could cover her up.

'Looks like you know what you're doing, Vinnie,' Norah said with a smirk. 'You been practising?'

Ignoring Norah, I held up the gown and Rachel poked her arms through. The gown did up with little ties at the back, and she leant forward, her face resting against my chest, so I could tie them.

'Want help getting the pants off?' Norah said, the joking tone gone from her voice. 'You don't want to get the bandages wet.'

Rachel clung to me and shook her head. Her face was strained, a hard line standing out between her brows. I guessed she was starting to feel the pain.

'We're right, thanks, Norah.' I pulled the gown down over her so she wouldn't be exposed there in the bright light, and then I reached underneath and carefully inched the pants down her legs. She sucked in a breath when I reached her feet, and real slow I edged the wet material over the bandages. The pants were singed around the hem but I threw them over a chair. Gem might still want them for something, I didn't know.

Stranded there on the bed, the oversized hospital gown slipping off one shoulder, Rachel looked like a victim of war. Someone you might see in a newspaper, blown up in some distant land. She reached for me, like a child, holding out both arms as though I might lift her up and carry her away. I felt if I bent towards her she'd never let me go, so I just pulled the blankets up over her. She squeezed her eyes shut, dropping her arms onto the mattress.

'Vincent.' My name was a moan, deep and low.

'Norah, I think the pain is starting to kick in,' I called out, and my voice seemed to echo against the yellowish tiles, hard and tight sounding. Norah walked back across the room.

'Is it hurting, love?' Norah leant down and looked into

Rachel's eyes. 'I'll rig up this line, then you'll get some help with the pain.'

I watched as the needle pierced the skin, two red dots building around the hole. Norah had the IV up and running real quick, and after she'd got the fluids going through, she emptied a syringe straight in and Rachel's body relaxed.

'Morphine,' Norah said. 'That'll take the edge off.'

In minutes Rachel's breathing slowed, and she looked up at me through hazy eyes. I fished my phone from my pocket to check the time. It was already seven—I'd missed footy training completely. Two missed calls from Gem. Didn't even hear them. She'd be wondering where I was.

'Rach,' I reached out and tugged her gown back onto her shoulder, 'I got to go now.'

She stared at me, long and unblinking, as though I spoke a foreign language.

'I got to get home to Gem, okay?'

Nodding, slow and unsteady, Rachel turned her head away and closed her eyes.

'She'll go to sleep now, I reckon,' Norah said, watching me through narrow eyes. 'We'll move her up to the main ward a bit later.' She tapped the pencil impatiently against her clipboard as I hovered there at the end of the bed.

'Right, well. I better head off then.' My voice came out loud in the quiet of the room.

'Okay, Vinnie.' Norah stuck the end of the pencil in her mouth.

I reached down and scooped the scorched pants off the chair. 'They're Gem's—I better take them with me, I guess.'

'Sure, they'd probably just end up in the bin here.'

Rachel was already sleeping. Looking at her face, I fought the urge to cup her cheek in my palm.

'Well, tell her I said bye, alright?'

Norah scribbled something on Rachel's chart.

'Yeah, sure. I'll tell her.'

As I walked from the room I could feel Norah's gaze on my back. Outside it was dark, and the concrete steps of the hospital were lit up in a bright semicircle. I reached into my pocket for my ciggies before I remembered they weren't there. My hands shook in the light, my bare arms and chest breaking out in a rush of goose bumps. I stood there a minute, cold and stunned, and then stumbled out of the rim of light, back towards my truck.

I I

I was worried when Dad didn't come home that night. He usually rushes up the hill after work on a Thursday and then races around to get everything together for training. Sometimes I head down to the footy fields with him, but lately I'd been staying home a bit, doing assignments and stuff for school.

Dad didn't come home, even after training would have finished. I had dinner all ready, enough for Rachel too, but he didn't come and he didn't call. I started to stress. He's all over the place, my dad, but he always calls when he can't make it, when he's stuck somewhere. I started thinking something had gone real wrong. He didn't answer his phone. I rang Marie, and my Aunty Jan. I even rang Gran, though I knew Dad would hate that.

When he walked in, no shirt, all covered in black marks, my jeans hung over his shoulder, I started to cry. Couldn't help it. I'm not a blubbering type of girl, but he looked like he'd stepped out of *Mad Max* or something, and I guess I was already wound up. He wrapped his arm round me and then shuffled inside to find his bag of pot. He didn't say anything. Usually he wouldn't smoke dope in front of me, though I know he smokes, and he knows I know. But that night he just rolled the joint right then and there, lit it up, had a few drags and then went to find a jumper. I watched him smoke, sitting on the cane chair outside in the dark.

'Where's Rachel?' I asked when he'd taken the final pull on his joint and stubbed it out in the jar lid.

'Hospital.' It was more like a grunt than a word.

'What happened?'

'Stepped in a fire, burnt her feet.'

'Did you take her to work with you?'

I wondered if the joint had given him any relief. I'd smoked a few times with my friends and it didn't seem to me like it took the edge off reality. When I'm stoned I feel even more aware of the tiniest things. I don't much like it, to be honest.

'Yep.'

'Is she alright?'

'Yeah, I guess. In shock they said. Let's hope they can keep her there a while, so she can get it together.' He leant back in his chair and closed his eyes.

'Do you want me to heat up dinner?' I asked, thinking he looked about ready to collapse into bed.

'Nah, thanks, Gem. Maybe I'll take it as leftovers tomorrow, hey. Have to explain to the boss why I pissed off before he got there.' Dad rubbed his forehead with smudgy fingers. 'Guess he'll take that well.' He looked at me sideways, a tired smile. 'Sorry I didn't call you.' He stood up. 'I should have, but fuck, it was all a fucking disaster. Left the doona and shit out at work too. Hope it doesn't piss down in the night.'

Squeezing my arm, he passed me, heading for bed, and I could smell the pot he'd smoked lingering behind him, sort of sour and fermented. I looked around the kitchen. My homework books piled messily on the table. Dirty saucepans in the sink. The three cold bowls of pasta still on the kitchen bench. My stomach rumbled, and I picked up a bowl and started to eat.

The house was cold. I guess I should have lit the fire earlier, but I'd been too busy worrying to think. We've got one of those potbelly type things that doesn't really do much, but the house is so small that by evening it does make a difference. Dad usually lights it when he gets home from training the boys, and then by bedtime it's toasty. The cold made my hands and feet hurt, and I knew there was nothing for it but to go to bed. I filled up a hot water bottle and turned out the lights.

Tucked up beneath the covers I shivered for a while, then the heat of the hot water bottle spread out and I felt myself relax. Before sleeping I like to lie there and think about things. My secret thoughts, I call them.

That night I thought about Dad, and how sometimes it seemed like he just bounced from the centre of one disaster to another. He runs around so much, always on the go, but maybe if he just stopped and stood still then all the disasters might stop too. I didn't know what happened in his mind—if there was room in all his hurrying to have secret thoughts like mine. It was hard to tell how much he thought things through.

I wondered if I would be like Dad when I was an adult. Tucked up under my doona, I didn't feel like the world revolved around me enough to inspire those kinds of situations. I just seemed to be caught on the fringes of other people's little earthquakes.

Earthquakes started me thinking about my mum, about when I might see her next. I thought about how quickly my excitement faded when I saw her. How quickly she could disappoint me. How, in the first moment she walked in the door, I wanted to hug her, and at the same time I wanted to pull her hair and see her cry out, loudly, in pain. I never did either, just stood by and watched her, waiting for the fall. I waited for her to do the thing she always did. The thing that always made me hate her.

Last time she came she stole my piggy bank. She's always on the lookout for cash. It was old, with just a bit of change in it. I didn't care about the coins, but my gran gave me that piggy bank when I was real little, and even though it was a kiddy thing, I loved it. And she probably didn't know that, she's never around enough to know much about me, but she's my mum, and a part of me thinks she *should*. Dad was so angry, he went completely nuts. We drove around town looking for her, him swearing like mad, but she was long gone.

Thinking about Mum got me sad, so I tried thinking about something else instead, and Dave's face loomed up in my mind. He hadn't talked to me since Dad and Marie had their last fight.

At school I sit with my friends under C block, it's our hang, and he sits across the way against a big brick wall with a bunch of boys. When we sit and eat lunch sometimes I can feel him watching me. Sometimes I sneak a peek, but he always looks away and makes a joke with his friends, and I think I'm just imagining it, all of it. I don't know if I should try and talk to him or not. I think about what I could say, and it all seems so dumb, so meaningless, and I can't bear for him to look at me as though I'm not worth anything.

Sometimes boys can do that. Treat you real different when there are other kids around to what they do when it's just the two of you. Once a boy spent all night talking to me at a party and kissed me goodbye at the end—a real kiss, long, with his tongue inside my mouth. And then on Monday at school he acted like he'd never seen me before, even though everyone had seen us together. I was shocked, and it made me feel so small, as though inside I was shrivelling up. I wanted to tell Dad about it, but I thought he might freak out about the kiss.

He seems real cool about stuff, my dad, but sometimes he explodes about things like that. I can just imagine what he was doing at my age. Running riot, going wild. My gran says he was always in trouble, but when it's me it's different. Kind of unfair, I reckon, specially since there are so many things I'd like to ask.

That night under my doona, I wondered if Dave was mad at me, if Marie had said something that made him think there was something wrong with me. Sometimes I think there is something wrong with me. No-one else seems to have these secret thoughts winding around inside their minds. Everyone carries on in any given minute as though it all makes sense—the world, school, home. But when you think about it, nothing really does.

For the next week I watched my dad, watched him get quieter and quieter. He lost his smokes on the day Rachel got burnt, and he didn't have enough cash to buy more. It took me a little while to notice, and then I saw that he couldn't sit down, kept pulling at invisible threads on his clothes. Jittery and fidgeting like mad.

'No fags,' he said, when he caught me staring at his shaky hands.

I went into my room and came back out with the ten bucks Gran had given me the week before.

'I don't want to borrow your money to buy cigarettes, Gem.' He tried not to look at the note in my fingers. 'You know I don't do that shit.'

'Take it,' I said, hating to see him suffer. 'You'll pay me back. I know.'

He reached out and took the money, fingers trembling, and shoved it in his pocket.

'You're something, aren't you?' He smiled at me, reaching out to ruffle my hair like I was just a little kid, but I could see he was ashamed. 'Payday soon,' he said, jumping in the truck and heading off to buy the ciggies from the late-night servo in town. I watched him speed off, dust rising out behind the truck. It was lonely standing in the doorway, staring out at the dark, waiting for him to come back.

He didn't go and visit her—Rachel. In hospital with her burnt feet. I kept waiting for him to say he was just dropping in to check on her, but as the week dragged on he said less and less.

We saw Marie down at footy on the weekend and she walked right by, looking the other way and laughing with her girlfriend. I felt Dad tense beside me, felt his whole arm go hard, but he just raised his head and stared at her, watched her walk by. Dave and Pete didn't come near us either, though they waved from the other side of the footy field. Dad and I lifted our hands in unison, a sad little salute.

I didn't know what had happened, but it felt different down at footy. It felt like everyone stood just that little bit further away from us. The kids were fine, racing around tripping each other up and acting like a bunch of rowdy monkeys, but the mums, they stood back, just a little, and I wondered why. I guessed Marie had said some stuff, that some rumour about Rachel had gone round. When I watched those women, how easily they could turn away from you, it reminded me of the bitchy girls at school, and I wondered how it could be that nothing ever changed.

Dad was smoking pot every night. He sat outside in the cold, but I could smell it drifting in my bedroom window. He stopped shaving, and his beard came on real thick and black. He looked like a pirate, all woolly and wild. I hadn't seen him like that for a long while, and I wondered who he missed the most. Who it was all about. Marie or the broken girl.

I thought about Rachel a lot, with her plastered arm and burnt feet. Those red strands of hair in among all the black. The leaking milk. Her big, dark, watching eyes. I wondered if she was okay. By the end of the next week Dad was barely speaking at all, except to swear now and then, and I was starting to fret.

'Dad, have you heard anything about Rachel, how she's going?' I finally asked on the Friday night while he ate the dinner I had cooked, not even seeming to see it. The lines around his eyes were deeper, etched in somehow.

'Scott called me and told me she was alright, healing up and everything.' He looked out the window, avoiding my eyes.

'Why don't you go see her, Dad?' The words dropped from my mouth and hung there in the air.

Dad looked at me, his face hard, but kind of hurt too.

'Gem, she's not my fucking problem.' He was angry, his voice loud. 'I found her on the road. Fuck, it's not like she's my girlfriend. I don't even know her.'

I looked down at my plate, gathering up some peas and spooning them into my mouth. I like the way peas pop, the way you can almost hear the noise they make when they burst. A bit like berries. I thought about Dad's words. It seemed strange to me that he should get so mad if he really didn't care. Swallowing the peas, I watched his face.

'When she turned up you took care of her like she mattered, like she mattered to you.' I didn't know how to explain what was bothering me. 'And then you pretend she doesn't exist. I don't get that.'

Dad took a long swig of the beer he was drinking and then placed the bottle carefully on the table.

'I'm just trying to do the right thing.'

He looked real tired, sitting there under the bare globe.

'She needs help, Gem, proper help. You know? From someone trained.'

'But don't you want to check yourself she's okay? What if she thinks you don't care?'

'I'll tell you something.' He picked up his beer again. 'Things might seem a certain way from the outside, but when you're inside them, it's a whole different angle.' Dad's voice had softened, it was almost gentle. 'People always think they know what's going on, but they don't know shit, because they weren't there, and they never could be.'

'Dad, what are you talking about?' I started thinking maybe he was drunk, but he'd only had that one beer.

'Everybody thinks I fucked her, because she's young and pretty and, you know, vulnerable.' The last word tripped from his mouth—I don't think I'd ever heard him say it. 'Even you, Gem, and you're my daughter.' He pushed his plate away. 'And Marie goes around stirring up her shit, bad-mouthing me to anyone that'll listen, and there's nothing I can do but sit here and take it, cause I can never explain it, and the more I'd try the weirder it'd sound.'

'Explain what?' I felt closer than I ever had to finding out something I wanted very badly to know.

'I dunno, Gem. Just what it was like.' Dad stood up and put his plate in the sink. 'Thanks for making dinner, mate.' He swiped the back of his hand across his mouth. 'Think I'd better hit the sack. Got to get up early to watch the boys play. You going to come down to footy in the morning?'

I stared at him, willing him to tell me something more, something about how things worked.

'Gem?'

'Nah, I'm helping Gran in the morning, remember? You got to drop me there on the way to footy, then I'm staying at Mel's. You said you'd pick me up Sunday from her house. You said it was okay.'

'Yeah? Fuck, I forgot. Okay, night, sweetheart.'

'Night, Dad.'

He went straight to bed, no pot left to get him through. I stacked the dishes in the sink and thought some more about Rachel and her sad lost-baby song. How she was that day I came home from school and found her, crooning on the grass by the side of the road. I could feel the echoes of the tune, her song, in my mind, and that night, hugging the hot water bottle to my belly, I couldn't sleep.

Dad said he was trying to do the right thing, but I couldn't help thinking that maybe he was wrong.

12

After the boys' game had finished, I was collecting up the balls and stuff when Scott wandered over carrying his littlest fella. He must have come down to the fields to watch one of his lot play, but I could see right away that he had something on his mind. His little boy, George, looked at me from his father's arms and said, 'Da ball, da ball,' in his husky baby voice. He held out his tiny arms, and I handed him a footy.

'Already got him playing, eh, Scott?' I said, thinking about what it must be like to have a family like Scott's—a smart wife who wasn't going nowhere, and four healthy kids.

Scott straightened the beanie on the baby's head, jigging him a little on his hip.

'Yeah, even little Evie, though there isn't a girls' team.'

'Get her playing with the boys.'

'You growing a beard, Vince?' Scott said with a grin. 'Looking pretty rough, mate. Going for the bushranger thing?'

I laughed and rubbed my beard.

'A wog Ned Kelly.' I said. 'Nah, I just hate shaving. Guess I'll have to soon or I'll be scaring all the kids.'

The baby dropped the ball and I reached out and caught it, a reflex. George giggled, like he thought it was a game just for him. I handed it back and he dropped it again, watching my eyes real close, his face one big smile. Scott caught the ball this time and then held it tucked up under his arm.

'Mate, I just thought I'd tell you,' he looked back over his shoulder, checking who was around, 'Rachel's ex's turned up. He came to the hospital yesterday, kicked up a bit of a stink about seeing her. Nurses wouldn't let him in.'

'She's still up there?'

'The hospital's not busy, lots of empty beds. They'll keep her up there a few more days yet—if she needs it.'

I didn't want to care about Rachel's bloke, but I did. George stretched out his little body and grabbed the ball under Scott's arm.

'I just wanted to let you know. She asks about you, Vince, every time I go in there. She asks when you're coming.'

'The fucking bastard.' I could feel my teeth gritting. 'Where is he? Where's he staying?'

'I don't know.' Scott looked at me then, his eyes sharp. 'You've got to remember, Vince, it was his baby too. There'll be a service on Monday. He's organising it.' Scott held up his hand like he knew what I was going to say. 'He's the father— even if he's a fuck-up, he has the right to be involved.'

'Soon as you lay hands on a woman like that, you got no rights.' My fists were tight balls. I tried to let go of the tension but my voice sounded rocky. 'You saw her, you saw what he done. Frankie wouldn't be dead now and she wouldn't be fucked up like she is if he had just behaved like a fucking human being.'

Scott put George down. 'Georgie, just go kick the ball over there for a sec,' he said to the baby, and George stood a minute and then toddled off.

'Sorry, I should watch my fucking mouth.'

Scott shook his head, watching the little fella wander away with the ball.

'The weird thing is, the bloke, he wasn't young. Norah reckons he was older than me. Old enough to be her dad, she said. They were all a bit shocked. Pretty girl like that.'

'Fuck, an old geezer, older than us?' I was surprised. I don't know what I'd expected but I guess it wasn't that.

'Yeah. Anyway, just thought I'd let you know. I don't think Rachel's heard he's around yet, cause the girls wouldn't even let him on the ward. The funeral should have been earlier but it got postponed because of the autopsy. You should go.'

'Mmm,' I said, thinking. 'What happened with the autopsy? What'd they find?'

'No sign of abuse. He was just so little, the impact of the crash sort of crushed him even though he was in his baby seat. Internal injuries. I guess she'd been there a while before you came. It might have helped if they'd gotten to him earlier, I don't know.'

'Fucking country roads. It's a real bad corner.'

'Yeah, that's what the police said when they interviewed her in the hospital. No blood alcohol reading, no sign of culpable driving. It was just a dangerous corner late at night. Makes you think, doesn't it?'

I could see Scott watching George. The baby toppled backwards onto his bum. He sat holding the ball, trying to stand back up, then he threw the ball away and slowly pushed himself to his feet to trail the bouncing footy.

'About what you've got,' I said, putting my hand on Scott's shoulder, knowing what he was thinking, thinking it too. George wobbled towards us with the ball.

'Yeah. Too right.' Scott picked up George again in one arm.

'Thanks, matey,' I said, taking the footy from the baby's outstretched hands.

'Give me a call if you want to grab a beer later.' Scott adjusted the baby in against his body. 'Might be able to wrangle an hour or two away from the crew.'

'Okay,' I said, knowing that I wouldn't. 'See ya, George,' I called out as Scott walked away, and I could see his little hand fluttering above his dad's shoulder, waving.

I knew then that I should have gone and seen her. Rachel. That it was wrong to leave her alone. I thought of how she'd reached out for me in the hospital bed, how she'd held up two arms like a baby, expecting me to lift her up. I'd walked away, and I couldn't help feeling it was like walking away from George. Like leaving a baby stranded, lost and alone.

I rocked up to the hospital after the footy, thought I'd just walk up the hill from town. Picked a bunch of flowers on the way, but when I got there, standing on the front steps, they seemed a bit sad. Saggy and withered, hanging down from my fingers. I chucked them in the bin on my way inside.

The lady at the front desk didn't look up when I came in, and after a minute I cleared my throat. She peered at her screen for one more long second and then glanced up.

'Yes, can I help you?'

'I'm here to see Rachel. I know she's not taking visitors, but tell her it's Vincent.' I ran my hand across the top of the desk, waiting for the woman to move.

'I'm sorry, sir,' she said, looking at me over the top of her glasses. 'We checked her out this morning.'

'What?' The hairs on the back of my neck stood up.

'She wanted to go, and we can't keep a patient here against their will. Anyway, the doctor discharged her, she was looking much better.'

'What about her feet?'

'She was walking around.'

'Where'd she go?' I asked, feeling a sinking inside me.

'I'm afraid she didn't say where she was headed.'

'Did someone pick her up?'

'To be honest, I'm not sure what happened. It's not my job to keep tabs on people. She wasn't with anyone when I saw her.'

I just stood there in the lobby, speechless, wondering what to do. I hadn't thought she might disappear exactly as she'd come. Just keep on running. The woman behind the desk turned back to the computer in front of her and started up typing. There was nothing to do but go home, so I pushed back through the swinging doors. The faded flowers poking up through the hole in the bin seemed even sadder on the way out.

I'd blown it, whatever it was.

I'd dropped Gem at my mum's house in the morning and when I got home the place was quiet. I did last night's dishes and brought the washing in from outside and sorted it there on the couch, trying to catch up on all the stuff I couldn't manage during the week. The clothes Rachel nicked from the woman in the hospital had gone through with all the rest, and I folded them and put them on the kitchen table. I could see that they were old ladies' clothes, large and pastel, like something my grandma would have worn. I wondered if the old lady missed them.

I scrubbed out the shower then, thinking of how dirty it had seemed with Rachel curled up on the bottom, her pale nakedness somehow unsoiled even with all the bruises. The weight of the week seemed to pull on me, and I thought fuck it, I'm going to bed.

Sometimes that happens on the weekends. All the early mornings seem to add up and just knock me flat, and I need to take an afternoon to sleep it off. My bedroom door was open, and I slipped inside, pulling my shirt over my head. I stopped then, arms up high, cause she was there. Asleep on my bed. Poking out from the doona were her feet, the dirty bandages unravelling, and beside the bed was a small black bag, plonked down like she'd just dropped it there and fallen straight into sleep. I stood looking down at her, my heart beating heavy in my chest, but she didn't stir. She was wrapped up in the doona like a caterpillar, as though she'd rolled around in it and gotten caught. On my pillow her face was peaceful, the bruises gone, her lips parted just a bit. After a minute I crawled in beside her, careful and quiet. I lay there, head propped on my arm, watching her breathe, staying still so I wouldn't wake her.

She was so beautiful, like no-one I had ever seen, skin smooth and clear like china. Her black lashes long and curled like a cartoon girl. And in those slow minutes before I fell asleep I wondered what it would feel like if she were mine.

The ringing cut into my sleep and I was up with the phone in my hand before I was really awake. It was my mum, her voice strong and loud down the line.

'Vinnie, what you up to this afternoon?'

'Aw, Mum, I was asleep. Hang on a sec.' My voice sounded dry, like I'd smoked too many ciggies. I could hear Mum moving through her house, rattling dishes on the sink.

'You washing up?' I asked, still trying to shake myself awake.

'Might as well do something, waiting for you to switch on.'

I walked into the kitchen and splashed some water onto my face.

'So, Vinnie, what you up to this afternoon? The real estate's coming on Monday to do an inspection, and I've got all those cardboard boxes out the back from the shop. I need you to do a dump run for me, love. Can't have those boxes lying around, real estate won't like it.'

My mum runs a craft shop up the coast a little way, one of those places jammed full of endless shit. She loves it, that shop's her life, but her place is always overflowing with rubbish, boxes and packaging and leftover stock that she can never seem to get rid of. I rubbed my beard, thinking, not sure what to do about Rachel.

'What's Gem doing?' I asked.

'I've got her sorting buttons. She's heading off to Melissa's a bit later.'

'Okay.'

I wanted to tell Gemma that Rachel was back, that she was alright, but I didn't much feel like discussing it with Mum.

'I heard you got that girl staying up there with you again,' my mum said, and I groaned silently to myself. 'Heard she's a pretty young thing, Vinnie. Why don't you bring her round so I can meet her.'

My mum's one of those women who always has her ear to the ground, as though she picks up messages from some secret tunnel system with openings that come up right inside my house. Not everyone's like that. Lots of people round here stick to themselves. Don't ask, don't tell. Tucked up in the bush, no neighbours peering in, who'd even know I was here? But whatever chain of information Mum's a part of, I learnt early that there's no escaping it.

'Yeah, Mum. I got a girl here, but she's just a friend.' I bent my head around the corner to check Rachel was still sleeping. 'She's had a bit of a rough time. Don't know if she'll be up for tea and bickies.'

'I heard about the baby. So sad, Vinnie. Bring her round, darling. Gem and I'll look after her while you do the dump run.'

I stood there in the kitchen wondering what to do. My mum likes a cup of tea and a good chat. I could just imagine her sitting Rachel down and telling her whole life story, all about my sister Hannah, the whole fucking lot. Then she'd get on to me and Gemma's mum, and how I never made anything of myself. How I'm still living in a broken-down shack, working my arse off and never getting anywhere. I wasn't sure Rachel was up for it.

'Mum, you can't fill her head with shit. She's still real wobbly. If I bring her round there you're going to have to go easy. Right?'

'Vinnie, I been dealing with your sister for years. Don't tell me I don't know how to handle someone on the edge.'

I sighed. My mother's burdens to bear.

'How is Hannah?' I thought of my sister, grown bloated and lumpy from years of medication.

'I went round there the other day. Same same. Drinking too much coffee, not talking much. I hate seeing her like that. Remember how beautiful she used to be?'

'Yeah, Mum.' I didn't want to get her started.

'So you'll come round soon?' I could hear her tap running,

she was rinsing the dishes.

'Okay, we'll come a bit later.'

'Dump closes at four, don't leave it too late.'

'I know, we'll be there soon.'

Hanging up the phone I jumped in the shower, and when I came out Rachel was awake, sitting at the table with a glass of water. She stood up, holding her glass in a shaking hand. She looked a bit startled seeing me in just a towel, all wet and dripping, and I wasn't sure how to make it alright.

'Vincent, I let myself in, I'm sorry—you weren't here.' It was the most normal sentence I'd heard her say.

'S'alright, Rach. I never lock the door.'

She looked so uncertain, standing there, sleep-crumpled and trembling.

'Listen, my mum wants me to go do a job for her. You want to come? I'll just get dressed.' I popped into the bedroom and threw on some clothes. When I came back out she was still standing there.

'You've grown a beard.'

'Nah, I've just been lazy.' I said. 'I keep forgetting to shave it off. Maybe I'll do it now, before we go.'

In the bathroom I rubbed the foggy mirror with a towel. She came and stood in the doorway.

'You eaten today, Rach?' I started soaping up my face.

'I ate breakfast at the hospital.'

I glanced over at her. She was wearing a skirt, sort of orangish and kind of sticking out, and a pretty blue top with flowers embroidered round the neck. The clothes looked somehow mismatched and not quite warm enough. She had a proper sling on this time, white and clean.

'You got some new clothes.' I looked down at her feet in the bandages. 'How're your feet?'

She sipped her water.

'Alright, they don't hurt.'

'Where'd you find the clothes?'

She watched my face underneath the white foam, as though she didn't quite trust it was me.

'Scott brought them in. His wife went through her old stuff, the stuff she didn't want anymore. He just came in and gave me the whole bag. Nice, hey?'

I pulled the razor across my cheeks in small jerky movements. The hair on my face was thick and tough.

'Yeah, that's real nice. You got any warm stuff?'

'I don't know. I just put on what was on top. I still don't have any undies though.' Her eyes darted sideways. 'No underwear, I mean.'

I tried not to think of her without undies, and leant down to splash water on my face, washing off the bits of soap. Out of the corner of my eye I could see her shift her weight from leg to leg.

'How'd you get here, Rach?' My voice was muffled by the water. 'You didn't hitch again, did you?'

'One of the nurses gave me a lift. I told her I was staying with you.'

Patting my face with the towel, I turned off the tap.

'That's good. It would have been shit to walk all the way up here on those feet.'

She moved out of the doorway as I came through.

'You want to come meet my mum? I have to take some stuff to the dump for her. Gem will be there too.' I picked up my ciggies and put them in my pocket. 'She'll give you some food, so we won't need to worry about lunch. You wait, she'll be plying you with fucking cookies and custard and God knows what else.'

'You don't mind?' She put her glass down on the table. She looked delicate and cold, and I wanted to throw a rug around her and snuggle her up.

'No, it's no worries. Let's find you a jumper, though. You just want to borrow one of mine?'

She shrugged, her mouth turning down at the edges.

'I can't feel anything, Vincent. Like I'm numb on the inside.'

I went into the bedroom and came out with a big fleecy navy-blue thing with a yellow stripe down the front. Thought it'd be easy to get over the plaster.

'Take the sling off, and we'll pop this on.'

I didn't know what to say to her about being numb, about losing Frankie. I didn't know what words to use. She pulled the sling over her head and I helped her put on the jumper. I felt better seeing her rugged up. Slipping the sling back on, I adjusted it around her plastered arm.

'Has the milk stopped?' I thought about the let-down, all her spraying milk.

She nodded and her eyes filled up with tears.

'Mostly,' she said, husky voiced. 'It's mostly gone, unless I squeeze a bit.'

'You seem good, Rach, you're coming good.' I squeezed her good arm through the fleece, softly, not sure if I should touch her, but wanting to.

'He came to the hospital. Paul did. The nurses thought I didn't hear.' She tucked her hair behind her ear, slowly, and then pulled it back out again. 'I couldn't stay there anymore, Vincent. Not when I knew he'd come to find me.' Her big dark eyes were scared, and I let go of her arm.

'It's okay. It'll be okay.' My words sounded empty hanging there in the air. I walked to the door and lit up a smoke. 'We'll go in a sec, just let me have a quick smoke.'

13

I was surprised when Gran told me that Rachel was back at home with Dad. She'd heard it from one of the nurses, must have had a phone call. Gran didn't make anything of it, but I knew she was worried. Worried about what kind of situation Dad had gotten himself into. She asked me a heap of questions about Rachel, but I didn't know anything. I wondered if Dad knew anything about her either.

We were sitting in the kitchen gasbagging when we heard the front door open.

'Shit, Mum, when you going to get rid of some of this crap?' Dad called out loudly from the hallway.

My gran collects stuff for her shop, and this morning there were all these bundles of zippers stacked on top of boxes of buttons, right there in the doorway. It was hard to even get in the door.

'Come on, Rach, it's just through here,' I heard him say, and when they walked in he was holding her hand.

'Vinnie,' Gran said, and reached out her arms to hug him. 'You look all fresh and clean. Last I heard you were growing a beard.'

Dad rubbed his chin. 'Word gets around, hey?' He looked at me and smiled. 'How many buttons you sorted?' he asked.

'Thousands,' I said. 'She's got me working.'

He laughed, looking around the kitchen. He knew I'd spent the morning cooking.

'This must be Rachel.' Gran leant forward, putting an arm around her and kissing her cheek. 'I'm Jude, Vinnie's mum.'

Rachel stood stiff in Gran's embrace, still hanging onto Dad's hand. I smiled at her from the table, but she didn't smile back. It was hard to tell if she even knew who I was. She had bandages on her feet, and I wondered if they still hurt.

'Now, I'll put on the kettle for tea, and we can have a little cuppa,' Gran said, turning around.

Dad opened a cupboard, pulling out a jar of biscuits. Taking a handful, he popped one in his mouth. He's always like a big kid when he comes back here, munching away at sweets like he's been starved.

'Now listen, feed her up real good while I'm gone. We haven't had lunch yet.'

Gran smiled. 'You like carrot cake, Rachel? Gem and I made a nice carrot cake this morning. Better get you some real food first though, eh?'

Dad pulled open the fridge. 'What you got in here?' He poked his head inside. 'You got pie. What type of pie is it?'

'A sort of Mexican thing, beans and tomato. I was experimenting. A recipe from *Woman's Day.*'

'Looks good.' He cut a piece and bit into it.

'Vinnie, I'll heat it up first, love.' Gran looked at Rachel and winked. 'He's always in such a hurry, my boy. Even when he was just a little fella I could never get him to sit down and eat ...'

'Mum,' Dad interrupted her, leaning back against the sink, 'you said you wouldn't fill her head with crap.'

Gran picked up the dish and put it in the microwave. Her back was turned, but she kept talking over her shoulder. '... well, he was the sixth, see. And by then, it's like anything goes.'

I could see Rachel was barely listening. She was watching Dad like she was afraid of him leaving.

'Okay, the boxes are out the back, aren't they, Mum?' He headed for the back door. 'I better get going or I'll miss the dump.' Pausing, he stood a second watching Rachel.

'You going to be right, Rach? I'll be gone about an hour. I'm taking Mum's trailer, so I'll have to drop it back. I won't forget you. And Gem's here too. Don't worry.'

Rachel lifted her good arm to hold the plaster, but she didn't say a word.

'See you soon,' he said and closed the door behind him.

The microwave beeped, three loud electronic pulses, and Gran took out the plate and cut up the pie.

'Here, love. Sit down at the table and eat.' Her voice was gentle and she pulled out a chair. 'I was going to make tea, but you look like you could do with a big glass of milk. Milk and pie, how's that sound?' Gran filled two glasses with milk.

Rachel nodded, taking a glass in her hand.

'I'm sorry about your baby, love,' Gran said straight up, sliding her hand across the table towards Rachel. 'I know there's nothing I can say, there aren't words for it, but I just wanted you to know that I feel for you.'

Rachel's lips turned up, a sad flickering smile. I could see her hands gripping the glass.

'I've had trouble enough in my life, but I never lost a child. Touch wood.' Gran knocked her knuckles against the wood of the table.

It was hard for me to imagine what it would feel like to lose your baby. I wished I could say something to Rachel about it, something like Gran just said, but my mind was blank. I couldn't think—nothing seemed right. I put my fork down and picked up my milk.

'Now tell me, where you from, love?' Gran motioned to me to keep eating, and she cut into her own piece of pie with a fork. 'Such a pretty girl, you look like you're part something.'

Rachel pierced her pie. She was awkward, one-handed.

'I was born in Sydney.' Her voice was quiet, her body still. 'I grew up with just my mum. I never knew my dad. My mum was Australian, I mean of Irish descent.' She looked up at Gran and me across the table.

'My dad was something, from somewhere I guess, but my mum wouldn't talk about it, so I don't know.'

Gran put her fork down, both elbows on the table.

'You never met your dad? And your mum never told you about him?'

Rachel nodded and some pastry came away on her fork.

'Mum was travelling when she got pregnant with me. Through India and the Middle East, Morocco, all over really. I think something bad must have happened, but she would never say.' The words seemed to slip from her mouth. 'She was open about most things, but about that she'd just say, "I can't tell you. Please don't ask me." I didn't like to see her sad, so after a while I stopped asking.'

It was more than I'd ever heard Rachel say. That's the thing about my gran, she can get anyone talking. She just asks all the questions that other people might think are rude, but it never sounds rude when she does it. Rachel pulled at the pie, her plate slithering on the table. I wanted to help her, but I wasn't sure how.

'Why don't I cut it up for you?' Gran said, taking her plate. She divided the pie into bite-sized pieces with her fork, attacking it roughly, her head down.

'Well, that's quite a story, love. Did you hear that, Gem?'

I could only nod. I felt a bit silly. I'd have to be deaf not to hear.

'So where's your mum now?' Gran kept up the questions, putting her glasses on and peering across the table at Rachel again, as though to get a proper look.

'She died. Breast cancer. When I was nineteen.' Rachel moved the pieces of pie around her plate, not eating.

'That must have been real hard,' Gran said, and I could see she was thinking. 'What about your grandparents? Your mum's parents. Where are they?'

'I don't know. When Mum came back pregnant they cut her off. They were religious or something. They lived in Queensland somewhere, I think.'

'Well, I bet they're regretting it now.' Gran shook her head, taking another bite of pie. 'I mean, family is family. How do you know they've never tried to find you?'

Rachel sipped her milk and then smiled across at me like I'd suddenly come into focus. Her face looked different, as though talking had made her less tight. 'We moved around so much when I was a kid. I guess it would have been hard, even if they tried.' Her voice was getting stronger. 'I always got the impression that they were country people, like way out in the bush. That's why Mum chose Sydney, cause she knew they'd have trouble finding her. But I don't know.' Rachel stopped and rubbed her nose. 'I could just be making stuff up. Some-times it's hard to tell what's real and what's not. Mum never said much about it.'

'What was your mum's name?'

'Sarah.'

'What was the family name?'

'Jones.'

'Know your grandparents' first names?'

'No. Mum never said.'

Gran watched her a minute, and then nudged Rachel's plate of pie. 'Come on, eat up, love, so we can get to the carrot cake.'

I watched Rachel chew the pie. She looked like she was struggling to swallow.

'What about your bloke?'

Rachel went completely still. 'I had to take Frankie,' she whispered. 'He said he loved him, but that's what he said about me.'

Gran's head tilted sideways. I could see her mind ticking. Rachel looked away, tracing a flower on the tablecloth with her good hand.

'Sometimes it didn't feel like love,' she said softly, not glancing up.

There was something private about what she'd said. Something I didn't understand. Gran didn't say anything for a while.

'Well, my Vinnie now, he's a good boy,' Gran said after a bit, patting Rachel's hand. 'Always been a little wild, but sweet-natured all the same. He'll look out for you.' Gran stood up, walking to the kettle. 'I'll just make us some tea and we can go and drink it in the sunroom, with the cake.'

In the sunroom, Rachel seemed stunned by the brightness. There was something I was missing, but I wasn't sure what. She took a sip of the tea, though it was still boiling hot.

'Let it cool down a bit, love,' Gran said, watching her carefully.

The sunlight shone on Rachel's bare shins showing up all the black hairs. She crossed her bandaged feet, and uncrossed them again. I wondered how much older she was than me.

'Looks like you need to change those bandages, you got any more?' Gran leant down to cut the cake.

'The nurse gave me some when I left.'

'Make sure you get Vinnie to help you change them when you get back. You don't want to get an infection, darling.'

Rachel put her teacup on the small cane table. She looked a little wobbly.

'My Vinnie's got a real knack for helping things mend. When he was a little fella he used to bring home all these creatures. You know, a kitten with a bunged-up foot, a possum that had been rounded up by some dogs. Once he even brought a turtle that he found on the road, been hit by a car. He'd spend hours coaxing them back to life.' Gran bit into her piece of cake and wiped the crumbs from her lips. 'Well, sometimes they were too far gone, you know. I don't think the poor turtle made it.'

I loved it when Gran talked about Dad. He never told me anything about when he was a kid. It's not that he was secretive, he just didn't seem to think stories were important. He was always on the go, in the middle of things.

'I always figured he should have been a vet, or even just a nurse maybe.' Gran ran a hand through her hair. 'All my other boys went into the trades, and that would have been good for Vinnie too, I guess, but Hannah was in trouble by then. His dad had passed away, God rest him, and I needed help looking after her.'

'Who's Hannah?' Rachel asked, looking down at the piece of cake on her lap but not touching it.

For my gran, everything always came back to Hannah.

'Vinnie's sister, my fifth. She's got the schizophrenia.' Gran edged forward on her chair. 'I couldn't keep a hold of her, but I sent Vinnie on her trail. He watched over her for me. Tried to keep her out of trouble.'

Rachel picked up her cup of tea, but she didn't take a sip.

'It broke my heart, it did, watching my girl.' Gran cut another slice of cake and lifted it to her mouth. 'She's on the drugs now, but it's not much of a life.'

I could see the teacup begin to shake in Rachel's hands. Leaning forward, I took it from her.

'You right, love?' Gran asked. 'You look a little peaky.'

'I'm hot,' Rachel said, and the plate slipped from her knees.

'Maybe you should come out of the sun.' Gran stood up and grasped her arm. 'Come on, come and sit in the lounge.'

I picked up the spilled cake and followed them into the lounge room. Rachel sat in the corner of the couch, enveloped in Gran's lacy white cushions. She looked tiny, like a child. I sat down on the chair opposite, not sure where to be. I spent a lot of time with Gran when I was little. After Mum left. I was used to sitting in on her conversations, to just listening, but with Rachel it felt a little strange. She wasn't one of my gran's gossipy friends. She was the girl Dad found on the road, the girl with the dead baby.

Gran took the spilled cake from me and shuffled from the room, coming back in with some glasses of water.

'Those are my grandkids,' she said, pointing to a crowd of family photographs on the sideboard beside the couch. 'I got thirteen all up now, counting my Gem here.' Gran smiled at me when she said that. She never treated me differently cause I wasn't really Dad's.

Rachel glanced at the photos but looked away real fast. There was a photo of my grandad on the opposite wall.

'That's Vinnie's dad, back before I met him. He was an Italian. Well, I met him here, but his family was from Italy.' Gran watched Rachel's face. 'Spitting image, isn't he?'

Rachel nodded, but she looked pale.

'I'm just going to pack up the cake, girls. I'll be back in a sec, you just sit and relax,' Gran said, putting the glasses of water on the sideboard.

I tried to think of something to say—something normal, something easy—but nothing came out. Rachel pressed at her eyes, squeezing her forehead, pinching the skin together in her fingertips. I looked from her to the pictures. All the babies. My cousins. Toothy grins, dimples and plump baby arms. Girls with pink ribbons in their hair. The photographs seemed to loom over Rachel. Even I could feel their weight.

Suddenly she scrambled up, stumbling blindly from the room into the kitchen. I followed her, not knowing what else to do.

'Rachel?' Gran pulled her head out from the fridge.

Standing there, Rachel shuddered.

'Love?'

Shaking her head, Rachel closed her eyes, her good hand clenching at her side.

'Darling.' Gran stepped up and wrapped her arms around her. 'It just hurts. It just bloody hurts.' She squeezed Rachel, pulling her in against her chest. 'I know. You can let it all go. Just let it all go.'

I stood there watching.

Rachel seemed to burst. A strange distant sound, an awful choking sob.

'What else can you do but cry, love. What else can you do but cry.'

I was stranded. Sometimes it feels like your feet are glued to the floor, and every step you take would cost more than you could pay. Watching her and Gran was like that. I dragged myself outside, and headed off into town to find Mel.

14

She vomited last night when I got her home from Mum's. In the garden, by the door. I gave her a drink and washed her face, then just wrapped her up and took her to bed, but I couldn't sleep. In the night she sweated beside me. I could see the sheen of moisture on her face even in the pale moonlight. She murmured and groaned, and I lay there listening to her dreams.

In the morning I left her sleeping, sneaking outside to have a ciggie. Even before I lit up the breath left my mouth in white puffs, and in the distance ice glistened on the grass. It was quiet there in the half dark, and after my ciggie I turned the hose on, careful not to wake her, and washed away the brown lumps. Bloody Mexican pie. I should have known Mum would be too much for her.

When I got back from the dump yesterday she was sitting at Mum's kitchen table, dull-eyed and blotchy-skinned. Not talking, just staring. Gem had already gone to Melissa's. Mum was fussing around preparing dinner, chattering away, telling her all about some lady down the road whose dog was sick or something.

'Rach, you alright, baby?' I asked, looking across at Mum. Rachel didn't answer.

'Mum, you said you'd go easy. What'd you fucking say?'

Motioning me out the back, Mum gave me one of her looks, and I followed her outside to the concrete patio.

'Vinnie, she's got to cry. She's got to talk about it. That's grieving. That's how you get over things.' Mum's arms were crossed, high and stubborn.

'But when I dropped her here she was fine and now look at her.'

'She's just lost her baby, Vincent. What was it—two weeks ago? Nothing is going to be fine. You can't make that okay.' Mum rubbed my arm then, scruffing me like I was a dog. 'She's got to get through it. You can't bypass pain, Vinnie. It's not healthy.'

When my mum talked like that, it just made me want to leave. Jump in the truck and drive off.

'Mum, I said she wasn't up for it. You didn't tell her about Hannah and all that shit too, did you?'

'Darling, I just got her talking. That's it.'

I glanced in the window at Rachel's slack face.

'You asked her anything yet, Vinnie?' Mum watched my face. 'Her bloke was hitting her, wasn't he?'

'She was bruised up pretty bad when I found her. But I thought it was from the crash.'

'She's tougher than she looks. Most women in her situation don't get this far.'

'What do you mean?' I was getting impatient.

'My cousin Faye in Sydney, her husband used to hit her

all the time. She was always hiding the bruises. She thought she deserved it. Nothing we said ever made any difference.'

'That was then. It's different now, Mum.'

'Is it?'

I wanted to get going but I could see that Mum wasn't finished.

'It used to shock me how much everyone ended up despising her. You'd think it would work the other way. In the supermarket people would turn away, as though she'd clobbered herself.' Mum took off her glasses and cleaned them on her shirt. 'Her husband was such a bloody bastard.'

'Why didn't she leave him then?' I didn't want to be talking about some long-lost relative in Sydney.

'That's what I'm saying. Something strange happens. Faye didn't start off believing she was worthless, but after she'd been knocked around for a while, that's how it seemed to her. She didn't want to leave. She was frightened too, I guess.'

'Rachel's shit-scared.'

'Good reason, I'd say.'

I reached for the door handle, ready to go.

'She doesn't have any family, you know, Vinnie. Her mum died of cancer. She's never met her dad.'

It was all coming out.

'What'd you do, ask her whole life story? Good gossip for down at the bowls club, Mum?'

'Shut up, Vinnie.' My mum doesn't take any shit. 'She's got grandparents up north somewhere. Bet they'd like to meet her—maybe I could try and track them down?'

'Fuck, Mum, it's none of your business, alright?'

'Put an ad in the *Woman's Day*?' She looked up at the guttering then, thinking. 'If I had a grandchild that I'd never met who needed help like this, I'd want to know about it.'

I wanted to tell her to keep out of it, but I could see she'd mulled it all over in her brain already.

Rachel opened the door then, steadying herself against the doorknob, eyes red and glassy.

'Vincent?' she sounded confused, like she'd lost her footing.

'You ready to go home?' I shook my head at Mum, willing her to let it go.

Rachel's hand was clammy in mine as I led her out to the truck.

'I feel sick, Vincent. I feel really sick.'

Back at home, she caught me by surprise when she vomited, covering her mouth with her hand and staggering sideways. She leant against the doorframe and bent over the garden in the dark. I could see her body jolt, as though she'd been hit from behind.

I had run out of words by then, so I just held her shoulder till it was finished and took her inside.

In the morning I had to go and pick up Gem. I was thinking of just leaving Rachel there, letting her sleep, but she got up just when I was ready to go. The orange skirt was crushed, sticking out all wrong, her pretty shirt falling open at the top from where she'd pulled the sling on. It had been two weeks since I'd first seen her, but in that short time she seemed to have shrunk. I swear the flesh was falling off her bones. She was dark under the eyes, and I knew she needed to eat. I thought maybe porridge, the way my mum makes it. I took off the sling and got out the plastic bag and rubber band, wrapping up her arm.

'Go have a shower, Rach,' I said, handing her a towel. 'Wake yourself up.'

I tipped oats and milk into a saucepan, and hunted out some cinnamon in the cupboard. Must have been there for years, but it smelled alright. I peeled an apple, chopping it into the steaming mix. I could hear the shower running, and I hoped she was managing. I tried not to think about her tussling with her clothes.

When she came out the porridge was ready, cooling on the table. She stood in the doorway, wrapped in the towel, black hair plastered down around her.

'My feet are weeping,' she said, lifting up her foot. 'I don't think I'm supposed to get them wet.'

'Shit.' I hadn't even thought of her feet. 'Do you have something to put on them?'

'The nurse gave me heaps. Pads and bandages, that salty water stuff.' I could see she didn't want to step forward. 'It's all in the bag.'

I left her standing there while I got the bag from the

bedroom. In the front pocket was a stash of white hospital gear, bits and pieces all in separate plastic packets.

'Okay, just come and sit here, Rach,' I said, moving the piles of clothes off the couch. 'Prop your feet up and I'll check them out.'

She stepped gingerly towards me, as if she was worried about her feet wetting the floor.

'Sweetheart, nothing you do to this place is going to make any difference. Look around you, the damage's been done.'

She glanced about then, as though she'd never noticed it before.

'It's a funny house, Vincent.'

'Yeah, it's an old dog of a place, but you get used to it.'

When she sat down on the couch, her towel slid up her thighs. I didn't fancy sitting there and staring up the length of her legs, trying to pull my eyes away from the black shadow of her crotch, so I went and got the doona and threw it over her.

'It's a bit cold,' I said, picking up her feet and propping them on a pillow.

'Is it?' She shivered, wide-eyed.

Her soles had peeled off in circles, the fresh skin underneath shiny and pink. One big toe was weeping, the wound not quite healed. I broke open a container of saline solution, squirting it over her feet. I didn't know exactly what to do with all those different packets of stuff, but I'd seen enough footy injuries to have some idea. There were little specks of dirt from the floor caught in the crevices of her skin, and I tried to wash them away. After patting her soles dry with a cotton ball, I smeared on the lotion.

Wrapping the clean white bandages around her feet felt comforting, somehow. I glanced up at her face when I was finished, and she stared at me from the corner of the couch as though watching a TV screen, as though my activity was something distant and removed from her.

'I have to go pick up Gem,' I said, glancing at the wonky clock on the wall. She sat up, pushing the doona away.

'Can I come?'

Her towel was coming apart down the front. I could see glimpses of pale skin.

'I'll have a look in here for some warmer clothes.' Turning my eyes away, I picked up her bag and tipped the clothes out onto the end of the couch, sifting through them for some long pants or something. There were a couple of pairs of jeans, scuffed at the knees. I grabbed one and pulled out a long-sleeved skivvy thing that looked warm, then I shoved the rest back in the bag. Laying the clothes over the arm of the couch, I picked up my fags.

'You get dressed while I have a smoke, Rach,' I said. 'Then you'd better eat up the porridge. It'll be cool by now.'

I stood out the back smoking, but I could see her through the window, pulling the skivvy on one-handed. I waited a little bit before I went back in.

'They're a bit big.' Her eyes were large, and she held up

her top. Her pale belly glowed flat and smooth, and the top of the jeans gaped loose around her. One jagged purple line, like lightning, marked her skin. A stretch mark, I guessed. Apart from that one mark, it was hard to imagine she'd grown a baby in there. Some women are like that. My wife was the same.

'I'll find a belt. And some socks to cover the bandages. You sit down and eat.'

I hunted around in Gem's room for a belt, but the only thing I could find was one she'd had since she was little, with Mickey Mouse heads all the way round. It looked smallish but I thought it might fit. I grabbed some big warm socks on my way back out.

Sitting at the table, she was spooning up porridge in her wrong-handed way. I couldn't see her getting the socks on, so I crouched down and pulled them over the bandages. Hanging the belt over the back of her chair, I stood up to get her a drink of water.

I was restless watching her eat the porridge. Gathering up the piles of sorted washing, I put them away. I got the broom and swept the floor, a tangled bunch of her fallen hair sitting lightly on top of the small pile of dust. Must have been there since the week before. I swept it into the dustpan then chucked it out the window into the garden.

'You ready?' I asked when I saw her slowing down.

She pushed the bowl away and stood, picking up the belt in her free hand.

'Frankie had a Mickey Mouse shirt. I bought it at an op shop.'

I watched her rub her thumb across the cartoon face.

'You don't have to wear it, Rach. I'll find something else if you want.'

'It's okay,' she said, lifting her skivvy, holding it up between her teeth. She couldn't see over the plastered arm, so I reached out and threaded the belt through the loops. It did up on the last hole, and I guess it did the job. She looked like a kid—oversized clothes and a Mickey Mouse belt, hair a big tangle. I wondered just how old she was, but I didn't ask.

She picked up my packet of ciggies then and put them in my shirt pocket.

'Ready to go?' she said, as if it was just a normal day.

'Ready to go.'

When Gem saw me at Melissa's door she gathered up her stuff, quick smart, and didn't want to linger. In the truck we were all squashed up in the front, Rachel's hard plaster pressing against my side.

I drove through town cause there was something I wanted to get. It was a Sunday but I knew Best & Less would be open till twelve. Town is pretty empty on Sundays, no-one wandering down the streets. The café on the corner stays open nowadays, so there were a couple of people around. I pulled up out the front of Best & Less and stepped from the truck. Gemma and Rachel tumbled out the other side.

'What are we doing, Dad?' Gem asked, looking at me across the windshield.

'We just need to get something quickly from in here.' I pointed towards the shop. 'Come on.'

Rachel stood there looking lost and I motioned her inside. The girls followed me in and I scouted around. The shop was chock-a-block with clothes, all bundled together on the racks. I just wanted to go in quick and then get out. Against the side wall were the tubs of women's undies. Two bucks each. I guess I looked like a bit of a sicko checking them out. My mum usually bought Gem a few packs for Christmas so I'd never done it before.

'Dad, what are you doing?' Gem whispered behind me.

I turned round to her, rubbing my hand through my hair.

'She's got no undies, Gem. I mean, she had nothing, and Scott's wife gave her a bag of clothes but no underwear.' I picked up a pair, looking at the tag. 'They all look the same, which ones should I get?'

Gem stood beside me looking down at the pants all laid out in neat piles.

'Just get normal ones, Dad.' She reached her hand into the tub.

'Which ones are normal ones?'

'Okay.' Gem held up a pair. 'Like these ones. Not with silly patterns, not too huge. Bikini-style.'

I picked up a plain pair, a pale sort of yellow.

'Low-cut or high-cut?' I checked out the tag again.

'What size?'

'Just get some plain colours, size ten to twelve. I doubt she'll be fussy, Dad.' Gemma looked around. 'Hey, where is she?'

The shop was so packed with stuff we couldn't see her. Gemma and I walked out to the front section so we could look down the aisles. There was movement at the checkout and I glanced over. I recognised the girl standing there behind the white counter. I hadn't noticed her on the way in. She waved and I knew I'd have to say hello.

'You go find her, Gem.' I nodded towards the aisles of clothes. 'I'll grab the pants and meet you at the counter.'

I went back to the tubs of underwear, bundled up a small pile of the pants that said ten to twelve, and then I headed up to the checkout.

'How you been?' I said, putting the pants on the counter. 'Haven't seen you in a while.'

Vicki was a girl I slept with on and off a few years back. Just two lonely people I spose, there was never anything in it. She was raising a couple of kids on her own, doing it tough. Nice girl. It ended easy. Mutual and all that.

'Good, Vinnie, good. Been working, you know?'

I could see her looking down at the pants, but I couldn't bring myself to slide them across into her hands.

'How're the kids?' I said, hoping Gem would come in a minute.

'They're fine. Don't usually work today, it's dead quiet. Liz called in sick, so Mum's got them for me.'

'Liking school?' I glanced behind me, looking for the girls.

'The little one does, he's totally keen. But Timmy, you know, he's trouble.'

Gem came up then, and I knew straight away something was wrong.

'What?'

'Dad, you better come see.'

I followed Gem down the rows of clothes. Past the women's and men's, all the way down the back till we hit the kids' section.

Rachel was kneeling on the floor in front of the baby clothes, looking down. She'd collected up a bunch of stuff— one of those stretchy baby-suit things, a warm jacket, some tiny blue socks and a baby beanie. She'd laid them out on the floor in the shape of a baby, putting the jacket on over the arms of the suit, pulling the socks over the feet. Gem and I stood behind her, staring down over her shoulder, but Rachel didn't turn round. She didn't try to hide what she was doing. She pulled a baby blanket off the shelf and threw it over the baby clothes, carefully uncovering the beanie again, bending over and touching the carpet where his face would have been.

I stood there and watched, not knowing what to do. Looking across at Gem, I saw she was struggling not to cry.

Vicki walked up behind us, poking around my back to see.

'Vinnie, what's she doing?' Vicki's voice was hard. 'Are you going to buy that stuff, miss?'

Rachel turned around then.

'Frankie was this size.' Her voice was strangely solid. 'See how little his feet were?'

'I know, Rach. I'm sorry.' I knelt down there beside her.

'Vinnie?' Behind me Vicki sounded pissed off. 'What's she doing? She's got to get those clothes off the floor.'

Rachel ignored her like she wasn't even there.

'We've got to put this stuff back now, Rach.' I lifted the baby blanket and started folding it up.

'No.' Holding out her plastered arm to ward me off, Rachel ran her fingers along the length of the suit, gently feeling for what was not there.

'What's going on?' I could hear Vicki's tapping foot. 'Vinnie?'

'Her baby died,' Gemma spoke quietly to Vicki behind me.

I could feel Vicki take a step backwards. 'Aw shit, really?'

'Yeah.' Gem scuffed her feet against the carpet and I turned back around.

'Look, Vicki, can you ring those pants through?' I handed her twenty dollars. 'I think I might have to pick her up. Carry her out. What you think, Gem?'

Gemma shrugged, but then she came and knelt beside me.

'Rachel?' she said, but Rachel just closed her eyes. 'Do you want to buy one of these things? To take home with us?' Gem scraped around in her pocket. 'I got ten dollars. We could probably buy the socks and the hat.' Gem leant down and checked the prices. 'Dad, give us two bucks and we can buy them both. What do you reckon, Rach?'

Rachel opened her eyes. 'The socks and the hat?' she said, turning to us.

'Yeah.'

Gem reached out and pulled the tiny blue socks free of the suit, soft and slow, and then handed them to Rachel. They sat in her open palm, and real careful I picked up the beanie and laid it on top. Rachel's hand trembled, and we stared at the quivering baby hat, the three of us all kneeling there together.

'Come on, let's go pay,' Gem said softly, and Rachel staggered slowly up.

The girls walked towards the counter, and I hung the rest of the baby clothes back on the racks. They were so small and light in my hands. A hollowness seemed to slide inside me, and for that instant I wished the shop had a back door I could just slip out and disappear through. But I could hear Gem's high sweet voice talking with Vicki, and I drifted down the front towards them.

Gemma handed over her ten bucks and Vicki already had my twenty. We stood there looking down at the counter as Vicki scanned our things through. She put the undies into a plastic bag, but she picked up the little socks and hat, stretching the scrunched socks flat, and placed them back onto Rachel's upturned hand. Then she opened the cash register and got me the change. Handing it over she said, 'Sorry about the baby, Vinnie. I hadn't heard. I didn't even know you'd broken it off with Marie.'

The meaning of what she said hit me kind of slow. I opened my mouth to say something. To say, 'No, no, it wasn't *my* baby,' but suddenly it all seemed useless. How to explain, what to say. I just nodded and handed the change to Gem, nudging the girls out of the shop.

'I don't need it, Dad,' Gemma said outside on the footpath, trying to hand the money back.

'Just take it, Gem.' I lifted my palms, the plastic bag of underpants swaying from round my wrist.

Rachel stood on the concrete, still staring down at the tiny clothes. Gem looked across at her, putting the money in her pocket.

'Come on, Rach, let's go home.' Gem's voice was soft, coaxing, as if Rachel was a sleepy toddler struggling to wake up. My girl walked over to the truck and opened the door.

'You hop in first,' she called out, waving Rachel over, 'and I'll climb in after you.'

'Frankie's dead,' Rachel said, not moving.

Gemma glanced at me, and suddenly she looked real tired and young. Like she'd somehow reached her limit.

'I'll get her, mate.' I lit up a ciggie, holding it between my lips. Wrapping an arm around Rachel's shoulder, I led her to the car. She seemed to drag her feet, and when I lifted her inside, I felt her resistance against my palms.

'Honey,' I said, 'it's a Sunday, there's nowhere else to go but home.'

She seemed to sag there in the front seat, looking down at the clothes in her hand. Gem got in after her, and I stood there a minute on the kerb, puffing on my fag.

'Dad, we're waiting,' Gem said, and I saw then she was a little shadowed under the eyes.

'You right?' I asked, watching her carefully for the first time. 'You tired or something?'

'I stayed up a bit late with Mel, that's all.'

She looked away from me, and she only does that when she's embarrassed or lying. I threw my ciggie in the gutter, squashing it with my shoe, thinking of all the things my girl could have done on a Saturday night. Sixteen years old and pretty as sunshine. I rubbed my face with my hands, trying to push away the images that suddenly flooded my mind. The truck felt cramped when I got in, cramped to bursting, the tiny socks and hat in Rachel's hand seeming to swell in size. As I turned the key in the ignition I wound down the window, trying to let the air in, wishing the cool wind would just blow all the sorrow away.

15

Some stuff happened at Mel's, some stuff I don't want Dad to know.

Mel's mum had headed off down to the pub for her night out, and we were just hanging out watching some daggy DVD when we heard a tapping at the window. It was Dave and this other guy from school, Matt. Mel looked at me sideways and kind of smirked, and I wondered if she'd known that they would come.

We walked out the back and said hi. The boys had a big bottle of Coke that they were sharing between them, swigging on it now and then. It was already half empty.

'What you drinking?' Mel said, flicking her hair.

She's an old friend, Mel. I've known her since primary school, and I love her, but she goes silly around boys. Can't stop playing with her hair, tying it up and then pulling it down. It's like a disorder. Obsessive compulsive or whatever it's called.

'Jim Beam,' Matt said, and handed her the bottle.

Mel grinned at me and then took a sip. I saw her trying not to pull a face. It tastes pretty crap that stuff. I've drunk it a couple of times before. She passed the bottle over to me and I held it a second, wondering if I really wanted to drink it.

Dave was watching me, so I lifted the bottle and had a gulp. It was even grosser than I remembered and I couldn't help wiping my lips afterwards. I handed the bottle to him, and he sipped it and then stood still, hugging it against his chest. After a minute of just standing there, Matt reached out for it again and Dave handed it over.

At Mel's the driveway curves around the back, so the boys stood out on the concrete while we sort of hung in the doorway. No-one said anything for a while longer, and then Matt started to laugh. Just at nothing, and he couldn't stop. It made us all a bit giggly.

'Fuckhead,' Dave said, holding in a laugh. 'He's pissed already.'

'Come on, guys, you got to catch up to me!' Matt handed the bottle back to Mel. 'Let's get fucked up.'

'Mum's down the pub,' Mel said, taking a swig. 'You guys want to come in?'

We stepped back from the doorway and the boys came through. The DVD was still playing and Mel walked across and turned it off. It was a sappy girly flick and I think she was embarrassed.

'You want to put on some music?' she asked the boys, handing me the bottle of Coke.

'What you got?' Dave asked in reply, but he kept looking at me.

'Heaps of stuff,' she said. 'It's all in my room. Gem will show you where.'

I felt my cheeks go hot and I took another sip of the Jim Beam, hoping Dave wouldn't see me blush. Even though I'd never been in this situation before—two boys, two girls and a bottle of Jim Beam and Coke—there was something familiar about it. Something sad and sick-feeling. I liked Dave, and I wanted him to like me, but I felt like walking into Mel's bedroom would be following a sort of script. Some path that had been laid down years before, maybe forever. Like I was part of an old bad movie, and I didn't want to be.

Dave reached out and took the bottle from me, and in that second I knew that we were friends. He smiled like he would have smiled if we were hanging out at home with Dad and Marie, kicking the ball around out back. At school he doesn't look at me like that, his eyes don't *see* me. All the air rushed out of me in that one second, and I turned and walked towards the bedroom.

Inside, the bed was all crumpled and there were piles of Mel's dirty clothes all over the floor. I kicked the squashed-up bundles towards her wardrobe and then headed over to the stacks of CDs.

'What do you feel like?' I said, leaning down to read the CD spines.

Dave stood clutching that bottle, looking at me.

'Dunno.' He sipped the drink. 'What do you like?'

I looked at all those CDs piled up beside the bed and had no idea what to choose.

'You have a look,' I said, and he came and stood beside

me. He handed me the drink and I had another sip.

'You don't like it, do you, Jim Beam?' His arm brushed against mine.

'Nup. It's fucking gross.'

I don't usually swear. When I was still real young I went to this friend's house down the road for dinner. She was a quiet kind of girl but I liked her. I think her family were Christians or something, but I didn't know anything about that then. We were all just sitting around the table talking, I can't remember what about, but afterwards the mum took me aside, gentle-like, and she said, 'Sweetheart, in this house we don't use words like that. In this house they're bad words,' and I didn't even know which words she meant, so she told me. They moved a few years later, but after that dinner I was careful. I started watching people real close to see what kind of words they used. Words are important, they tell people who you are.

Dave reached out and touched the piles of CDs, and I felt like I should say something more.

'I like it when your lips go numb, but. Then you know you're drunk.'

He looked across at me.

'Have another sip.'

Dave watched me drink from the bottle and I could see him staring at my lips. It made me feel self-conscious and I wiped them with the back of my hand. Passing the bottle back to him, I bent down to look at the CDs again. My lips were tingly. I didn't know if it was from the alcohol or Dave looking at them or what.

Dave sat down on the bed, and it sort of sprang against his weight. He bounced and let out a funny low laugh. When I turned around he looked like a little kid and I had to smile.

'Give it a go, Gem, it's real bouncy, like a tramp.'

He's drunk, I thought, but I sat down anyway. We bounced there together on the edge of the mattress, laughing a little, until the Coke started to bubble, and we stopped and sat still.

Dave slipped his hand up my sleeve and ran his finger down the length of my arm, real slow. I shivered, my whole body twitching. He smiled, and then he leant down and kissed me. His lips against mine felt big and strange. My mouth was a little numb, I think, and we both tasted of Jim Beam and Coke.

I always expect kisses to be like in the movies, where something magic happens, some explosion of electricity, but it's not like that. It's just two pairs of lips moving together, and if you think too much about it, it starts to feel weird.

Dave leant into me and I opened my mouth a little. I knew he wanted to press his tongue against mine, and when he did, he moaned, soft and low. I felt the sound inside him, felt it push against my body, but I didn't feel what he must have felt to make that sound. I didn't feel like moaning. I kissed him back, trying to reach for that feeling. I tried to let go of whatever it was I held back, to let that feeling come upon me, but nothing happened, and I wondered why. I felt numb,

numb all over, as though the lips that were kissing weren't even mine. It made me sad, like there was something lacking in me, and I pressed a little against him, to make up for what I didn't feel. He groaned softly and slipped his hand up the back of my shirt, fiddling with the clasp on my bra. I went still then cause I knew I didn't want that. I didn't want him to take my clothes off, to touch me like that. He stopped kissing me, and looked at me, close up.

'You alright?' His voice was strained.

'Yeah, I'm cool.' I didn't know how to say that I wasn't.

He dropped his hand from my bra and rubbed his fingers slowly up and down my back. I liked Dave, and I'd wanted to kiss him, but in those seconds sitting on Mel's bed with his hand on my back, I just wanted to run. I stood up and he handed me the bottle. I took a quick sip, not wanting to, and then I grabbed a CD off the top.

'This looks good,' I said, handing him back the drink. 'Let's go see what the others are doing.'

When we got out there, the other two were going at it on the couch. Not having sex, just going for it, limbs all tangled, and Dave and I stood there watching for a minute, not knowing where else to look.

'Come on, let's go sit outside,' I said, and walked out the back door. It was cool out there. I needed a jumper but I wasn't keen on heading back in just yet.

Dave and I sat side by side on the plastic chairs, just

staring at the back door, sipping now and then from the Coke bottle. There didn't seem to be anything much to say. After a while he reached out and took my hand, running his thumb against my fingers. I think I was drunk by then, cause I'd stopped even feeling the cold. I just sat there and stared at that door, letting Dave play around with my hand.

'I heard about that girl of your dad's. How she stepped into a fire,' he said finally, his voice a little slow. 'What a crazy chick, hey? She sounds worse than my mum.'

'She isn't my dad's girl.' My words came out all wonky sounding, and I pulled away my hand.

'Whatever. Don't get shitty, Gem. I just meant that girl that was at your house.'

'Her baby died.' My mind felt heavy, I didn't know what else to say about Rachel.

'I know. Wonder where the father is? I mean, he must be pretty pissed off that she crashed the car, kills his baby, and then shacks up with your old man.' His voice grabbed at me and I wanted to move away.

'It was an accident. She didn't kill the baby. You know what our road is like.'

'My mum says she'd have to've been driving pretty fast.'

'It was dark.' I kept staring at the door, hoping it would open.

'So what?'

I thought about Marie, and how she switched so easily. Nice and then not. It was all too much.

'Your mum won't even say hello to me now. She's intense, she goes nuts about everything.' I turned to look at him. His face had closed down, as though he was trying to lock me out. He took a long drink of the bottle. It was looking pretty empty now.

'Well, you'd know all about that, wouldn't you?'

'What?' I didn't know quite what he was trying to say.

'You'd know all about nutcases, wouldn't you? What about your mum? At least my mum's not a fucking junkie.'

Anger seemed to balloon inside me and I wanted to push him backwards, to push him off his chair.

'Fuck you,' was all I said, sitting there and staring at the door again.

'Come on, Gem.' His voice sounded slurred, and in that minute I hated him. 'Everyone knows your dad was fucking around with that girl even before she crashed the car.'

'Yeah?'

'Yeah, you must have known too.'

I looked back at him. His eyes were narrow and hard, like he hated me too.

'I'd never seen her before. She was just someone Dad found on the road,' I said. 'My dad might be some things but he's not a fucking liar.'

Dave leant towards me, bringing his face up close, as though he wanted to see if I was lying. I stared right back, daring him to say something. Suddenly he grabbed my shoulders and kissed me hard. My mouth was real numb, and there was a tearing feeling about it. His teeth knocked against mine, and I shoved him away and stood up, staggering backwards.

When I opened the door, the room inside was quiet. Mel sat on the couch, staring at the black screen of the TV. Matt looked at me and stood up, slipping his shoes on and fiddling a bit with his pants.

'Well ...' He turned back to Mel. There was something about his voice that sounded cold, cold but cocky. 'I'll see you at school then. Got to go. Where's Dave?'

'Out the back.' I pointed to the door, feeling myself sway on my feet. Matt leant over and kissed Mel on the cheek. She twisted her hair up, her face blank and grey.

'Okay, bye,' he said, and almost jogged out the door.

As soon as the door closed, Mel burst into tears. I sat down beside her, feeling a little queasy. She sobbed for a bit and I put my arm around her shoulder. The world seemed unsteady and I think I must have been gripping her too hard, because she pulled my arm away.

'You're pissed,' she said, laughing a little while she cried. 'You can't even sit up straight.'

My mouth didn't seem to want to form words.

'Gem.' She held my face between her hands, coming up close and forcing me to look right at her. 'I did it. I finally did it.'

I nodded. There were so many things I wanted to ask but the sick feeling inside me seemed to swell, and it was all I could do to stand up.

'I'm going to vomit.'

'Not here. Shit, get to the toilet.'

I could feel Mel behind me propelling me forward, and when we were in the bathroom my knees just gave way and I spewed everything up while Mel held my hair.

'How much did you drink?' Mel asked, flushing the toilet when it was finally over.

I couldn't answer.

'Did he kiss you? Matt told me he likes you.'

I looked up at her from the floor, her eager tear-puffed face looming above me.

'What was it like?' I asked, my throat feeling gritty.

Pulling her hair into a ponytail, Mel moved across and peered at herself in the mirror. 'Do I look different?'

I shook my head, she looked the same to me.

'It hurt, but everyone says it does the first time.' She tied her hair up with an elastic band. 'I'm glad I got it out of the way.' She pulled me up off the floor. 'Do you reckon that means we're going out?' Wide-eyed and hopeful, she was back to her normal self. 'Matt and me?'

I leant over the sink and splashed my face, gulping down some water.

'Did you talk about it?' I asked, wiping my face with a towel.

'We didn't talk about anything.' She turned on the shower, stripping off quickly and stepping in.

'I dunno, then. I don't get boys.' My voice came out all

dead-sounding in among the steam. I could see her washing herself behind the pink curtain, washing herself *there*, and I felt weak, like I needed to lie down.

Mel stepped out, reaching for a towel and wrapping herself up.

'Come on, Gem, let's go to bed before Mum gets home.'

We wandered back to Mel's bedroom, the floor seeming to lurch a little beneath me. Inside, Mel closed the door and dumped the towel on the floor, slipping on her pyjamas. Falling onto the bed, we climbed beneath her covers.

'I'm going to feel shit in the morning, aren't I?' I said into the darkness.

'You spewed, that means you already got it out. It's better to spew.'

It felt like my head was beginning to roll.

'Hey, Mel, did you use anything?'

'What? Oh, you mean ...' She turned over in the bed, her back towards me. 'Nah, you can't get pregnant the first time, everyone knows that.'

'You sure?' I thought about that for a while, wondering how possible it could be.

'Yeah,' she said sleepily. 'And he pulled out just in case. That's why I had to have a shower. It was all over me.'

My head was spinning. It was so messy. I closed my eyes and just tried breathing, careful and slow, and when I woke it was morning and Dad was knocking on the door, come to get me, with Rachel in tow.

Standing in the doorway, I felt like if Dad looked at me too hard he'd see everything ... Dave kissing me, the Jim Beam and Coke, Dave's teeth clunking against mine, Mel and Matt having sex on the couch, me throwing up in the toilet, everything. I just wanted to get out of Mel's before anyone woke up, before my dad stood there a while chatting with her mum, before it all came tumbling down on me in a big heap.

Sitting there, bouncing in Dad's truck, my stomach churned. I looked down at Rachel's hair hanging long against my arm, and I counted those red hairs in among the black— one, two, three, four, five, six, seven, eight, nine, ten, and then again—until my mind was clear of all the shit. Rachel might have been crazy, stepping into that fire, but I thought I knew why she'd done it.

And in a weird way, I thought maybe if I was her, I'd have done it too.

The funeral was held in a little wooden church on the outskirts of town. Scott called me the night before, trying to track down Rachel. He told me about the service, that the baby's father had arranged it, and I put Rachel on the phone so he could tell her. I wasn't sure that she'd want to go with him there, her fella. I walked away to let her talk, but when I came back in she was crying, crying but nodding, and I wondered what Scott had said. She wanted to go, so I called the boss and got a day off work. No-one can refuse you a day off for a funeral.

Later, when I snuck outside for a quick ciggie, Gem followed me, wanting to have a word. She reckoned we should wash Rachel's hair before the funeral, that she could smell it when we were all in the truck.

'She can't do it by herself, Dad. You need two hands to squeeze out the shampoo,' Gem said, not looking at my eyes.

I could see my daughter was right, but I wasn't sure how to go about it. How to even suggest it. In the end, Gem just knocked on the door the next morning when Rachel was in the shower, went in, and came back out a few minutes later saying, 'That's done. Easy.'

My daughter got up real early that morning, way before she needed to. She wanted to come to the funeral and I didn't see why not, steady as she is. After she washed Rachel's hair, Gemma sorted out some clothes for her, some things that weren't quite so raggedy, that didn't fit her all wrong. Then, while Rachel sipped juice, Gem brushed that long black hair, and braided it in two neat braids like a schoolgirl. I think my mum must have taught Gem how to braid, cause I sure as hell don't know. She's seen so much shit, my girl, but she's got such a sweetness to her. Sometimes I wonder how that can be, but mostly I just count my lucky stars.

Rachel looked exposed without the tangle of her hair spread round her, but she didn't seem to notice, spruced up and clean, sipping slowly at her juice. She wouldn't eat breakfast though, wouldn't even try.

When we arrived, there were a few cars already pulled up on the grass. I could see my mum's little hatchback, and I wondered for just a sec how she'd known to come. Gemma popped out the door and stood looking up at us. Rachel didn't want to get out of the truck, she was quivering so bad.

'You go in and find Gran,' I said to Gem, and she nodded, heading towards the open church doors. Rachel was still shaking and I sat there, not sure what to do. Truth is, I didn't much want to go inside either. After a few minutes I turned to her and said, 'Is it him? Or is it Frankie?'

She looked at me squarely, like she was going to speak, and then looked away.

'All of us are there. Me, Scott, Gem, my mum. He won't get you alone.'

She breathed in and out, sort of loud, and then opened the truck door with a jerk. She stepped down, and I jumped out my side and came round to her. I wanted to help her, but I didn't know whether to hold her free hand or put my arm around her. She looked a little weak at the knees. The church doors were open, calling us in, but she stood stiffly against the truck looking down at her feet.

'Come on, Rach, we got this far, let's go in.'

'You won't let him near me?'

I could see now she was real scared. Like she was holding back from climbing back up inside the truck.

'I'll stay with you. He won't get past me.'

'He'll try and talk, Vincent. He's good at talking. He'll try and *make it right*, but those conversations always come full circle until it's all my fault and I deserve it.' She looked up at me, and for the first time I could see she was angry too. 'Like I have to pay for all the pain in the world. And now Frankie's dead, so imagine the price.'

'He won't get to you.'

'I can't pay anymore. I've got nothing left.'

If she didn't want to go in, I wouldn't try and make her. Frankie was in there, in a coffin, and a priest would probably say some words. But I didn't know what that meant in the scheme of things.

'You want to go home, then?' I asked, watching her pale face.

She looked across to the church doors.

'Scott said this was my chance to say goodbye to Frankie.' Her bottom lip was trembling.

I could still see the baby's blue face, clear as day. His tiny mouth, and my finger inside, trying to clear his airway, as if I knew how.

'Come on,' was all I said, and she folded herself in against me, shaking and clinging hard to my shirt.

Up the front, the coffin was white and so small it was almost unbearable. I could only glance at it as I walked with her down the aisle of the church. We came through the pews slowly, cause she was wobbly on her feet and I was looking out for him, her fella. I half expected him to pounce. There was a bloke, I guessed it was him—Paul—sitting right there in front of the priest as though he'd done no wrong.

My mum and Gemma were sitting in the middle and Mum turned and motioned for us to sit with her. They'd been saving us seats, though the place was near empty. We passed Scott with a small group of nurses from the hospital. The ladies must have come across from work, they were all dressed in their blue uniforms, and some of them were already sniffing into their hankies. Scott held up his hand in a wave and I lifted my fingers, but Rachel didn't take her eyes from the coffin up the front.

I guided her into the pew, and she stumbled along and sat down. Mum wrapped an arm around her shoulder, smiling across at me, teary-eyed, a box of tissues on her lap. He turned then, the man at the front, his eyes darting around until they fell on Rachel. Stopping, dead still, he stared from me to her to Mum to Gemma and back again. I could see his mind going into overdrive trying to figure out who the fuck we were. Rachel's eyes didn't stray from the white coffin, almost as though she was in a trance, and I was glad cause it meant she didn't have to face the man eyeballing her.

The priest started talking, welcoming us and rambling a bit in that strange singsong voice they use. I was only half listening. Rachel's fella turned back around. The back of his neck was wrinkled like a man past his prime. He had two strips of greyish hair growing down the back of his neck, spiky and sparse, running down towards his collared shirt. He must have been nearly fifty.

The neck of a man, an older man.

A man who'd lain beside her in the darkness. Who'd touched her, who'd slipped within her, who'd made a baby grow inside her. Who'd bruised and battered this girl beside me until she'd taken their baby and run into the night. I stared at that neck but I couldn't bring those two men to-gether in my mind. Watching him, I began thinking of how easy it would be to crack him over the head from behind. To take him out, one whack to that creased old neck.

The priest began to sing an old churchy song in an unsure kind of way, and after the first verse my mum pitched in, and then Scott and the nurses behind us. The song seemed to wake Rachel from whatever place she'd slipped into and she gripped my forearm hard in her fingers. From the corner of my eye I could see the sweat breaking out on her face. The man looked back at us again, and Rachel turned her face aside, leaning in towards me. The song seemed to drone on forever, Rachel still gripping my arm, and I could feel myself begin to sweat too, as though the space in the church had suddenly gotten smaller and there was no more air to breathe.

'And now Frankie's father would like to say a few words.' The priest nodded to the man down the front, and he stood, a piece of paper trembling in his hands.

Stepping up to the podium, he glanced around, his face grey and tight. Most of us knew he'd been hurting her, and I reckon he knew we all knew.

'I wasn't sure who would be here today,' the man began. 'Frankie died on a stretch of road so far away from his home. I imagined maybe there would be no-one here to mark his passing.' He smoothed out the paper in his hands. 'I could have flown the body home, had him buried there, but I thought—he died here ... and I decided on cremation.'

He looked down at Rachel then, staring hard at her face.

'I haven't seen Frankie's mother since he died. I didn't get to discuss any of this with her.'

Rachel's arm shook against mine.

'In the hospital, you people ...' he pointed down the back '... you people kept her from me. What right did you have to do that?'

I could hear Scott and the ladies shuffling around, and Rachel pressed her face further in towards my neck.

'She left me in the night, taking my baby, and now he's dead, and you all sit there and pass judgement on me ...' he paused, eyes darting around, '... while she sits down there and she won't even look at me.'

I stood up then, Rachel staggering up beside me. I don't know what I thought I'd do, but I wasn't sitting there listening to him talk shit. My mum grabbed onto the back of my jeans, hissing, 'Vinnie.'

Nervously, the priest stepped forward. 'Maybe you'd like to read from your speech? About your son?' he suggested softly, though the microphone relayed his voice down the half-empty pews.

The man looked around, suddenly seeming a bit unsteady.

'Okay,' he said. 'Okay.' He looked down at his sheet. 'Frankie was a beautiful baby. He didn't cry much. When he was born he looked like a little old man, but then all the wrinkles went away and he was lovely.'

The man's hands shook. I sat back down, slowly, and Rachel sat beside me.

'Frankie loved the mornings, that's when he did his smiling.' Frankie's father wavered on the edge of tears, and I could feel Rachel begin to cry beside me. 'There's so many things Frankie never got to do.'

The priest stepped forward again, putting a hand on the

man's shoulder. I wasn't sure if he was offering him comfort or reminding him to stick to his sheet.

'I try not to think of all the smiles I'll miss now he's gone.'

Mum slipped an arm round Rachel and handed her a tissue. She murmured, rubbing Rachel's arm, and I wondered if it was some strange woman-knowledge that allowed my mum to shoulder all that sorrow as though it was the most natural of things. I looked across at Gemma, on the other side of Mum, and she smiled at me, a scared little smile.

'Rachel?' the man called out to her, but she wouldn't look up. 'You can't avoid me forever,' he said. 'You just can't.'

The priest motioned to the man that he should finish. Rachel's fella stared out round the church, looking at us all one by one. He didn't move from the podium, just stood there staring. The priest stepped closer and slowly he folded up his sheet and stood aside.

'Grief is a hard burden to bear, especially with the loss of one so small. Let us all join together now in prayer,' he said gently, looking down at us. 'The Lord is my shepherd, I shall not want ...'

Even I knew this one, and I mumbled the words under my breath.

Funerals are supposed to be good for us, a chance to say goodbye, but I didn't know if this one had done much for anyone. The prayer finished and I could feel Rachel straining beside me. I wanted to get out of there. I guessed she did too.

Mum blew her nose and then asked quietly, 'You want to

go to the crematorium, love?'

Rachel glanced up at her old man, standing there beside the priest, and then over at the small white box.

'It's enough,' she said, and stood. 'It's enough.' She looked like she might start running then and there, so we got up and moved along the pews with her, shuffling towards the doorway.

'Rachel!' His voice rang out behind us. 'Wait!'

She didn't stop, and I could hear his footsteps quicken. I turned and held out my arm.

'She doesn't want to see you, mate,' I said, and he stopped in front of me. Scott was moving out of his pew towards us, and behind me I heard Rachel scramble out the door with Mum and Gem.

'She doesn't want to see me?' He stepped back and I could see he was thinking of busting past. 'Tell me, how would you know what she wants?'

'I dunno, mate, maybe the fact that she just ran out the door.' I couldn't help saying it. Thinking of his big hands laying into her made the blood rush to my head. He looked like he was going to hit me, and then Scott was there, standing beside me, looking professional in his doctor's gear.

'Just leave it,' Scott said to him. 'She's not ready.'

'Not ready?' He looked disbelieving.

'You just go back down and talk some more to your priest. Get some absolution. Confess your sins.' The words fell from my mouth and then he was on me, but I didn't care.

I was just looking for an excuse to knock him flat. I took a fist to the face, and then I got him a good one. Scott sprang in between us, pushing us apart, and I could see the priest grappling with the fella from behind.

'Mate, she doesn't need this shit,' Scott shouted at me, and I felt the anger edge away. 'Don't you think she's had enough?'

I could feel the blood drip from my nose.

'Go on, Vince, get her out of here,' Scott said. 'I'll deal with him.'

I walked out those doors and when the sunshine hit me I could hardly see.

Mum came up with a box of tissues.

'You always had to be the hero, Vinnie, even when you were the littlest.'

'The guy's a fuckwit.' My nose was throbbing.

'Clean yourself up, you'll give her a fright.' She dabbed at my nose with a tissue. 'She's in the truck. Gem's got her. I tried to get her to come with me in my car, but she wouldn't.'

'We better go, then.' I grabbed the tissue and held it to my nose, turning towards the truck. 'Before he comes out.'

Mum gathered me up and gave me a quick hug. 'I'll drop Gem at school. You take Rachel some place nice, okay? You got the day off. Take her some place nice.'

I took her to the only nice place I knew. Down to the river. There's a spot where we used to go in summer when I was a kid, the local swimming hole, but I knew there'd be no-one there in the winter.

We sat on the sandy bank in the sun, me breaking up sticks and chucking them into the water. Rachel looked sad and small. Shoes kicked off, knees pulled up in front of her, she took up so little space.

'He'll find me.' Her voice was dull and low. 'Maybe not today, but soon. He knows I'm with you.'

I watched her, wondering what to say. Men like that didn't frighten me—taking out their frustrations on a woman, like spoilt kids. She must have been through some full-on shit to be so scared, but I knew I could handle him.

'He doesn't know no-one round here.' I figured it was best to play the whole thing down.

'I tried to get away once before, but he found me.' She ran her fingers through the brown pebbly sand, making patterns, hand prints, and then brushing them away.

'He won't get to you.'

I could have taken Rachel to the police, made a statement, but I'd seen that shit before with a girl I'd knocked around with in the past. Her bloke took the AVO as a challenge. It didn't keep him away. And what could the police do anyway? They hadn't done anything for her yet. Just laying low was her best bet. I could look after her.

'He won't have a fucking clue who to ask, and I don't even have an address. All my mail goes to a post box in town.' I thought about it for a bit. In a town like mine, people didn't give random strangers directions to other people's homes. Not if they didn't know their business. 'All anyone would say is—"Yeah, Vince, he lives up on the hill."'

She didn't answer, I don't know if she believed me. We sat there for a while, the sun warm on our heads. Then I stood up and wandered around a bit, gathering some sticks and leaves. Coming back, I dumped them down in front of her.

'Let's make a coupla boats,' I said, scratching my head. 'Then we can race them.'

I looked at the leaves on the ground, trying to figure out which one was best to start with. Me and my brothers used to spend hours making leaf boats, we were real experts. We'd send them downriver and see where they ended up. I grabbed a long canoe-shaped leaf which was the obvious candidate for the base, and I hunted around for a couple of twigs to make the mast. Rachel picked up a leaf, a roundish green one, and twisted the stem in her fingers.

'It'd make a good sail,' I said, nodding at her leaf. 'Might need a slightly bigger one though.'

She pushed herself onto her feet, and crouched there

looking at the leaves.

'Like this?' she said, holding up a wide green leaf.

'Bingo.'

She passed it over and I wove a twig through it.

'See that palm frond?' I asked, pointing to the edge of the water. 'You want to go grab it. We can tear it up and use it to hold the sail in place.'

She got up and walked over, just wearing socks. Her burns were nearly better but the skin was still pink and tender so it seemed better to keep her feet bandaged up. We'd have to change the bandages when we got home.

She picked up the end of the palm frond and dragged it over.

'See those ends? If you split them you can tear it down the centre and you end up with thin strips, bit like rope, then we'll be home and hosed.'

She lifted her sling off and chucked it a bit further up the bank. Holding the palm frond still by putting it between her plaster and her body, she worked at splitting the end with her fingernail. I let her do that and I hunted around for another canoe-shaped leaf. When I had one, I came and crouched there beside her.

'You finish this one,' I said, handing over the first boat.

She held it in her working hand, staring down at it.

'I don't know what to do.'

'Okay, well you hold it and I'll thread up the sail.' I tore the palm frond into a smaller strip and threaded it through the sail-leaf, tying it to the canoe-leaf at both ends. Then the sail was fixed sort of straight. 'There you go, what do you think?'

'Will it float?'

'Dunno, I reckon, maybe.' I picked up the next leaf and rigged up another sail. 'You ready to test them out?' I looked at her face in that bright midday sun. 'We'll take them up to the rapids.'

She stumbled up the river bank behind me, holding her boat, until we got to the spot where the river narrowed and the water flowed faster.

'Okay, here, just pop it in and we'll watch it go.'

Her socks were filthy now, covered in muddy sand, and when she leant over the water I was half afraid she'd tumble in, so I held the back of her shirt just in case. She let go of her boat and it drifted down into the flow. I dropped mine in too, and it bobbed along behind. We stood there watching them float away.

'He's gone,' she said, all quavery voiced.

I took her back along the river bank and we sat down again on the sand. She held her plaster, looking frail, and I sat there wondering what to say. It was warm in the winter sun, my head itched, and I took off my jumper and scrunched it up into a ball.

'Have a lie-down, Rach,' I said, patting my rolled-up jumper. 'Give the shoulder a rest.'

She didn't move, just stared out at the water in that wide-eyed way of hers. After a minute I lay back on the bank, leaving the jumper for her and screening my face from the sun. It felt good lying there, no work to be done and the sun beating down on me, warm and humming. I felt her come lie beside me, her back against my side and her head resting in the crook of my arm. The smell of shampoo came drifting up from her hair, and I smiled to myself thinking of how proud Gem had been to get her hair washed.

It was the smallest thing, lying there in the sun, but even then I knew it was something special. Time slowing, the leaves swaying overhead, a bird flapping by. And all the while her and me lying there like there was nothing else on earth. Slowly I felt her body soften, and then those twitches, sleep twitches, and even though my arm was stiff beneath her head I lay there, still and quiet, knowing that nothing lasts forever, but sort of hoping that it could.

I must have fallen asleep too, cause suddenly the sun was gone, and it was cool. She stirred beside me, sitting up and looking around, a bit dazed.

'It's beautiful here,' she said, sleepily. One of her braids was coming unravelled, the ends all soft and crumpled.

'Yeah, I used to come here a lot when I was a kid.' I scooped up my jumper and popped it back on. 'We should head home. Gem'll be wondering where we are.'

I reckon we'd both copped a bit of sun, cause she was pink-cheeked and my face stung a little. Then I remembered my nose, and lifted a hand to feel if it was swollen. It didn't seem too bad. She staggered up, still staring at the river, and then she turned back to me and smiled. In the quiet, her tummy rumbled, loud, like it was making a point.

'I think I'm hungry,' she said and laughed.

That laugh made me want to pick her up and squeeze her, but I just smiled and said, 'Let's go.'

Back at home Gem was waiting, all excited, a little furry bundle in her arms.

'Dad, I found it on the way home.' I could see her trying not to smile too wide. 'Can we keep it? Look at it, Dad, isn't it so cute?'

It was a tiny kitten, skinny and dirty.

'Gem, we can't take in every little critter you find on the road.' The exact words my mother always said to me came tumbling from my mouth. 'Cat food costs money, you know. Not to mention vets and the whole shebang.'

Gem's face fell and she turned away, cradling the little cat in her arms. Rachel stood in the doorway, not coming in, and I glanced over at her.

'Gem's found a little cat, Rach, come and have a look.'

Rachel didn't move, and I wondered what it was that bothered her.

'Look at it, Dad.' Gem held it up towards me. 'How can you say no?'

The kitten was sort of speckled orangey-black, dusty-looking, with uneven markings on its face.

'Gem, that is the ugliest kitten I have ever seen,' I said, laughing a little. 'It looks like something that's come out of a vacuum cleaner.'

The kitten squirmed and Gem started stroking it, real soft.

'Can we keep it, Dad?'

I couldn't bring myself to say an outright no, so I just said, 'We'll see,' and left it at that.

'Any dinner plans?' I asked Gem, thinking of Rachel's rumbling tummy. 'What we got in the fridge?'

'Dunno—some veggies, I think there's some mince in the fridge.'

'I'll make a pasta bake then.' Pasta bake was my specialty. 'We got some pasta shells, don't we? In the cupboard?'

'Yep. I think so,' Gem said, and headed into her room with the kitten.

It was late afternoon by then and getting dark. I knew I needed to change Rachel's bandages, but we were all pretty hungry.

'Rach, come sit on the couch.' I pulled her inside and threw a doona over her. 'I'm just going to get some wood to light up the potbelly, then I'll make some food, okay? Gem'll help you clean up your feet.'

She nodded, looking past me towards Gem's room. I opened Gem's door and asked her to deal with the bandages.

'Okay, Rach. It won't be too long,' I said, and disappeared out the door to get the firewood.

When we sat down to eat, it smelled good, and I realised it'd been ages since I'd cooked anything. Gem and I hoed in, but Rachel was still awkward with that left hand, so she was going slow.

'It's good,' she said, sipping at her glass of water.

'Yeah? Hospital food's pretty crap, hey?'

'I couldn't taste anything so it didn't matter.' She lifted another spoonful, a bit rickety, and put it in her mouth.

'You two are sunburnt,' Gem said, between mouthfuls.

I looked at Rachel's pink cheeks. 'Bit a colour. Do us good.'

Gemma looked from me to Rach and back again. I could see her thinking. She stopped chewing and gulped down some water.

'I think I need to do a wee,' Rachel said, pushing out her chair and wandering towards the bathroom.

Gemma waited until the bathroom door closed and then she said, 'She looks brighter, a bit. Maybe the funeral did her good.'

'It's probably just the sun. I took her down to the river after.'

'That's why you were late?'

'Yep.'

'I couldn't figure out how her socks could've gotten so dirty.' Gem smiled. 'They looked like she'd tramped through mud. I thought maybe she'd done another runner.'

'Nah, we just hung at the river.' I didn't want to tell my daughter we'd fallen asleep in the sun. 'How was school?'

'Alright.' Gem looked down, moving the pasta around in her bowl.

'Just alright?'

'It's school. What you expect?' She looked up at me sideways. 'Hey, Dad, I got an assignment due on Thursday, do you reckon I could have tomorrow off to do it? I haven't started yet and it's bigger than I thought it was going to be.'

My Gem's such a good girl, always heading off to school without complaining. I didn't see that she couldn't have a day off now and again if she wanted it.

'Sure, if you need to.' The bathroom door opened and Rachel came back out. 'Rachel can stay here with you then.'

It was two problems solved, cause I was wondering what to do about leaving Rachel by herself while I went to work. The kitten came tearing out of Gem's room then, sliding out across the floor in front of Rachel and straight underneath a cupboard.

'Oh, she's gotten a fright,' Gem said, getting up from the table to check. She crouched down and peered beneath the cupboard, calling, 'Freckles. Freckles.'

I tapped the table for Rachel to come back and sit down.

'Freckles?' I asked Gem, trying not to smile.

'Yep. Cute name, hey? cause she's so speckledy,' Gem called over her shoulder, stretching her arm underneath the cupboard and bringing the kitten back out. 'Now she's purring. Silly little bugger.'

Gem sat on the couch, cuddling up with the kitten, and

Rachel came back and sat at the table. She ate some more spoons of pasta, and I got up and started stacking up the sink. The house was warming up, the potbelly crackling away, and it felt cosy in there. After a few minutes Rachel brought her plate to the sink, all the food gone.

'You ate up everything,' I said, and I felt like patting her on the head.

'I want to help. Let me do the dishes.' She stood there beside me in my crummy kitchen, the bare globe bearing down on her pink face. I looked at her arm in the sling, wondering how she'd go.

'I'll wash. You stack them on the rack,' I said, thinking it was a compromise.

Her face came across all stubborn.

'That's not a real job.'

I had to smile.

'Yeah, alright,' I said. 'You wash, I'll dry.' I grabbed a tea towel off the back of a chair.

'No, you relax. I'll wash and then I'll stack,' Rachel said, nudging me away from the sink, and I went and sat beside Gem on the couch.

'You shouldn't let her do it, Dad. It'll take her all night,' Gem whispered to me. The kitten was purring loudly, the sound strangely deep for such a little thing.

'She's alright. It's better if she does stuff. She'll feel better.'

'You reckon?' Gem looked up from stroking the kitten.

'It's not good to get stuck all day with just your thoughts, not good for anyone.' I stifled a yawn, maybe I hadn't slept that long at the river after all.

Gem watched Rachel at the sink, struggling with the plates. 'When you going to get us a new telly, Dad?' she asked, looking back down at the cat.

Our telly blew a few months back and I was waiting to find one in good nick at the dump. Truth was, I didn't much like the sound of the blaring box. I liked life better without it.

'One'll come along soon, Gem.' I stretched out my legs towards the potbelly. 'No use buying something new when there's plenty of good ones lying around that no-one wants anymore.'

A fork clattered in the sink.

'Anyway, there's never anything on the fucking box. Never anything worth watching.'

'We could get cable. Mel's got cable.'

'So we could have even more channels of nothing but crap. Fifty-seven channels and nothing on.' I reached out and touched the kitten's head. It was a funny little thing, like a mishmash of every possible type of cat in one. 'Nah, Gem. It's not worth it.'

Gem yawned too and stood up.

'I made up a box for her in my room, a little bed. Newspaper for kitty litter. I might head off. The fire's making me sleepy.'

She said goodnight and then I stood up to help Rachel at the sink. She'd only done about half the dishes.

'Come on, I'll do the rest.'

She looked down at her arm, frowning, and then handed me the sponge. Wandering over to the bookshelf, she peered at the books.

'Quilting,' she said, pulling one out. 'Is this yours?'

I turned around to have a look.

'Nah, don't know where that came from. Probably Mum's. Not big on embroidery, me.'

She sat on the couch and flicked through the book, and I finished the dishes and cleared off the table. I did a final tidy-up and then said, 'Well, I'm off to bed, Rach.'

She looked up at me from her book.

'Gem's going to stay home tomorrow to do some school-work, but I'll be off early, okay?'

She nodded, her face kind of still and sad. I knew she probably wanted to come in the bedroom with me, but I didn't know where to draw the line.

'You keep reading if you like. Keep the lamp on and I'll just switch off the top light.'

I stood there a minute wondering if I should do some-thing more for her. The doona was there, all ready for her to pull up.

'Okay. Night then,' I said finally, switching off the light and heading from the warm kitchen to the cool of my room.

18

When she slipped in beside me in the middle of the night I wasn't really surprised. It was dark and I'd been asleep, but I was used to her now and it didn't startle me none. She laid her palm against my chest and smoothed it downwards, and straight away I knew the touch was different. I felt myself go hard and I put my hand on top of hers.

'What you doing, Rach?' I said, holding her hand still.

She pressed herself against me, pulling her fingers free.

'I just want to feel something,' she murmured against my neck. 'Please.'

I'd been in lots of strange scrapes with women, but never one like this. I didn't want to take advantage of her. I didn't want to be that man. But in those moments in the dark all the reasons I had for keeping her away seemed meaningless. I wanted to fuck her and she shifted against me, her hand sliding down and down. I turned my face and kissed her, slipping my arm around her back. She pressed away from me, tugging at her clothes.

'I want to *feel* it, Vincent,' she said. 'You help me.'

I pulled off her sling, and then her shirt.

'Pants.' She slid her jeans down one-handed. She was naked now, and she pushed herself against me.

'You too.' She leant across and kissed me. 'I want to feel you.'

I tugged off my trackies and she pushed her leg in between mine. Her bandaged foot nudged against my calves, and the hard plaster on her arm pressed against my chest. I couldn't see her face in the dark, but I rolled over, putting my arms on either side of her.

'You sure, Rach?' I held myself above her on my elbows, wishing I could see her.

'Vincent. I want to feel something.'

I leant into her, bending down to find her mouth. She wrapped her good arm round my neck, holding me tightly. It felt right, it felt easy, and slowly I pressed myself inside her. The darkness was heavy, magnetic even, as though we were pushed together by the night, sticking to each other. There was a weight in the air and my mind went blank. I was lost in the feel of her inside that blackness. I let go, swallowed up, the sounds of us lost in the air.

She pressed against my chest weakly with her plaster, and I stopped moving, the sound of her suddenly breaking in.

'Rach?' I was panting.

'Stop, please stop.' Her voice was sodden and quiet, and I realised she was crying.

'Fuck, what's wrong?'

I rolled off her and turned on the lamp. She shifted up so she was sitting, covering her eyes from the light. One of her braids had come undone and the black hair floated down across her breast. I stood up and got my ciggies, opening the window and lighting one up.

'Rach?' My fingers shook, and I dragged hard, pulling the smoke deep inside.

'I couldn't feel anything.' She swiped at her eyes. 'I just … couldn't feel it.'

Honest to God, I'd never had a woman say that to me before. My whole body went tight, as though I was coiling up inside. I wanted to jump out that window and run. As far away as I could get and then some.

'What do you mean?'

'It just feels like nothing. I want to feel it. I thought I would with you. I thought it would work.'

I looked at her, exposed in the lamplight, thoughts racing through my brain. Rachel sniffed and took her hand away from her face. The air from the window was cold, and I could see the goose bumps spring up across her nakedness, but she didn't pull up the covers.

'Did you feel it with him?' I forced myself to ask her.

She looked across at me, eyes big and liquid.

'Nup.' She shook her head. 'Never.'

The air rushed out of me, I was so fucking relieved.

'You never came, ever, when you slept with him?'

She shook her head again, the loose hair on one side fluttering against her body.

'Are you for real?'

'There's something wrong with me.' Her voice was soft,

and she pulled her knees up, hugging them against herself. 'That's what he always said.'

'Why'd you sleep with him, if you didn't feel it?' I glanced at the black sky outside the window, jaw clenched, not sure that I wanted to hear her answer. But when she spoke I couldn't help turning and listening.

'I just … I was so alone and he was there.' Her words were coming out jumbled. 'I didn't want it, I never wanted it, but I wasn't sure how to … not do it.'

Rachel looked down at the bed, avoiding my eyes, and I dragged hard on my cigarette, trying to block out the image of his old body bearing down on hers.

'It didn't feel good, but I thought that was cause I didn't want it.' She looked up at me then. 'And after he started hitting me I stopped feeling anything at all, so it didn't seem weird.'

I thought about what she was saying, trying to find a way to understand it.

'What about doing it to yourself?'

'What?'

'Masturbation.'

'What about it?' Her voice was muffled against her knees.

'Did you ever do it?' I sucked on my ciggie, watching her.

'Before my mum got sick, when I was younger.'

'Did you feel that?'

She looked up at me, her cheeks suddenly pink.

'Did you come?'

'Yeah, I think.'

'Well. It works alright then.' I stubbed out my ciggie and threw it out the window. 'You ever have a man go down on you?'

'What?' She looked away again. 'No.' She twisted the end of her braid round the fingers of her good hand. 'Paul never did anything like that.'

'What about anyone else?'

She kept on twisting her hair, not meeting my eyes.

'I've never been with anyone else.' She finally looked up. 'A boy stuck his finger inside me once on the school bus. It was awful. I didn't know what to do so I just sat there, but it didn't feel like anything except a finger.'

'Why didn't you move? Tell him to fuck off.' I thought of Gem, on the school bus every day, and I knew she'd never tell me about something like that.

'I just felt stuck, he had me trapped. I didn't want to call out, cause I didn't want everyone to know. Like it was me who was doing something wrong, even though I knew I wasn't.'

'Was he your boyfriend?'

'No. Just a guy from school. He wasn't a nice guy.' She pulled the elastic from the end of her hair and slipped it over her wrist. I stood in the cold air and watched her undo the second braid. Her hair stood up around her head, wild-looking and fluffy.

'It's all curly,' I said, and I moved to sit on the end of the bed.

'Crimped. I used to love it like this when I was little. Used to make my mum plait it the night before school so it'd be crimped the next day.'

I reached across and touched her hair, my hand still shaking. The hair felt sort of wiry and thick.

'Feels funny, doesn't it.' She ran her hand through it and smiled at me, a sad little smile. Her lips were puffy and red, and I wanted to touch them but I didn't.

'Rach.' I dropped my hand from her hair and sat it on her knee. 'Sweetheart, what do you want from me? Cause I'm not sure I can give it to you.'

She looked at me as though she had no secrets, her face bare.

'I just get so full, like I'm going to overflow. And then other times I can't feel anything at all. I just want to feel something that isn't this.' She cupped her breasts with her plastered arm. 'This space here where Frankie used to be.'

I sat there watching her cradle her breasts and wondering if I should try and touch her, if we should try the whole thing again. I was thinking maybe I could take it more slowly. Maybe I could help her feel it. I'd slept with my fair share of women in the past and none of them had said anything about not getting off. But I figured with a girl like her I needed to start from the beginning. Go back to the days when I was a young fella and didn't know if I was going to strike it lucky. When it was all just fooling around. When it was better to see a girl come than see nothing at all. I put my fingers round her ankle and squeezed.

'Tell me when you feel it,' I said, watching her face. I held her ankle soft at first, and then a little harder. She didn't even blink, so I gripped her hard, like I was checking the tyres of Gem's bike.

'I feel that,' she said, giving me her gap-toothed grin.

'What about this?' I asked and gripped her calf, squeezing it in my fingers.

'Yep, it hurts a little.'

'Okay.' I loosened my hold. 'We don't want it to hurt, just for you to *feel* it.' I slid my hand up and ran my knuckles along the curve behind her knee, and she shifted her legs a little.

'That's ticklish,' she said, putting her hand out to ward me off.

'Ticklish is good.'

I moved my hand up further, holding the inside of her thigh. She went still then, like she knew what I was about, but she didn't say anything. I just sat there holding her thigh, wondering if she wanted me to touch her. Hoping I wasn't like the boy on the school bus, pressing unwanted fingers into her. Then she parted her legs a little, and I heard her breath suck in, and I knew. I lay on my belly between her legs, ready to go down.

'Don't look,' she whispered, her eyes flicking away from mine.

'I have to look, else I won't be able to see what I'm doing.'

Her thighs trembled and I didn't hesitate then. I hadn't gone down on a girl in years, but I knew how to do it. It took a little while but in the end we got there, and I knew she'd felt it. She quivered and shuddered, made a sort of whimpering sound, and then reached out and pulled me up. I laid my head on her breasts and she hugged me real tight, holding me there, as though she hoped I might dissolve into her. I could feel her heartbeat skipping against my ear, loud and jumpy.

'Squeeze me,' she said. 'Squeeze me really hard.'

I lifted my body up and then wrapped my arms around her. Rolling her over, I squashed her in against me.

'I felt it, Vincent. It felt good.'

'Yeah?' I laughed against her neck. 'That's how it's supposed to feel, baby. That's the aim.'

I rolled back on top of her, my hardness pressing against her. I wanted so bad to be inside her, even just for a few seconds. I didn't push it, just lay still against her, but she opened her legs and kissed me, and this time we were right as rain. I don't know if she felt it, but she moved with me, and when it was finished she curled around me so tight I couldn't move. I couldn't even lean over to turn off the lamp. So that's how we slept, tangled up and entwined. The glow of the light upon us. I couldn't feel where I stopped and she began.

In the morning I pulled myself free. Careful not to wake her, I switched off the lamp and left straight for work. It was still dark out, the sky brightening slowly, pinkish and pretty. The truck jolted beneath me and the trees flicked past, green and cold. Mist hung on the road in thin patches. I wound down my mountain, crossed through town, and headed out towards the orchard. Towards the boys, Steve and Jimmy. Towards the chainsaw and the secateurs. Towards another day of my life, exactly like the rest.

All the way there I thought of what I'd done. Of the line that I'd crossed, and if I'd ever be able to cross back. I could smell Rachel against my skin, the scent wafting around me like the mist on the road, until I turned on my dodgy air-conditioning to try and clear the fog away. But it stayed with me during the day, even when the clouds darkened then opened, pissing down and soaking me right through. The scent of her trailing me as I worked, so that come home time I was knocked out with just the thought of seeing her face.

19

After that night at Mel's, Dave had done just what I thought he would, and totally ignored me at school on Monday. Matt did the same to Mel and she took it pretty hard. When I got back from the funeral Mel didn't say much, just looked quiet, and then right before she got on the bus she said, 'Why doesn't he like me, Gem?' trying so hard to keep from crying. I had no answer to that, but something about it ate away at me. I wished I could ask Dad about it, but that would mean telling him that Mel slept with Matt, and anyway, when I got home he wasn't there to ask.

That afternoon I saw something funny out the bus window, and since no-one was home I decided to go back to investigate. I'd thought it was some sort of animal—I was thinking a baby possum or something. The road curves around pretty sharp and I'd seen the thing on the side with the drop, so I walked along the edge, looking down at the tree trunks rising up. I got to the spot where I thought it'd been and it was still there. A kitten, halfway up a tree, stuck and meowing. Clambering a little way down the bank, I pulled it off the tree trunk. Baby things always make your heart go gaga, and when I picked it up it started purring straight away and crawled up against my neck. I wasn't far from our place, a five-minute walk maybe, and there was no sign of a mother cat. I reckoned it'd been dumped.

I wondered who would dump a kitten, especially beside a road where it was likely to get run over, but then I remembered how Dad told me that his father used to just break their necks at birth. Back then it was normal. No-one de-sexed their cats. They just twisted their necks as if it was nothing, as if they weren't even alive. It seemed strange to me that men who loved their wives and children could break a kitten's neck without a thought. Dad said something about it not being right to let all the kittens grow up and go feral, but I could see he didn't much like the idea of breaking the necks himself.

People always want to make bad things seem noble, as if wringing a kitten's neck is the *responsible* thing to do. But you just need to look at a kitten for a minute to see that it's wrong.

I wasn't leaving that kitten out there to starve or get squashed flat, so I brought her home. I knew Dad wouldn't like it but what could he say? It was a small thing in the scheme of things, a kitten. I'm only sixteen, but even I can see that Dad couldn't fault me for bringing home a baby cat, when he'd brought home a whole person, and an adult at that.

The day after the funeral I stayed home from school. I didn't lie when I told Dad I had an assignment to do, but the truth is that sometimes school just feels like too much. Rachel didn't get up till almost lunchtime. She came wandering out of Dad's bedroom, all dreamy and soft-looking. I was used to it now, her sleeping in there, though I didn't have a clue what it meant.

'You had a big sleep, Rach,' I said, and I couldn't help thinking of her the day before, breathing hard in Dad's truck. I'd sat with her outside the church, waiting for Dad. She was so pale, her breaths coming in short gasps. I'd tried to calm her down but she'd just shook her head. I'd wanted her to tell me about the father of the baby, about what happened, about why she was so scared. But I didn't ask.

The kitten was curled up on the table, and when Rachel walked into the room it stood and stretched. I tapped it on the nose with my pencil and it went cross-eyed trying to look. It was impossible not to laugh but Rachel just turned away.

'Do you want me to make you some brekky?' I wasn't sure how much she could do herself.

'It's okay,' she said, stepping over to the kitchen bench. 'Thanks, though.'

She spread butter awkwardly onto some bread and then stood staring out the window, slowly chewing. I didn't know if I should try to make conversation or not.

'There's a car coming,' she said, her body suddenly stiff. She looked around like she needed somewhere to hide, and I stood up behind her and peered through the curtains.

'It's my mum.' My teeth were clenching already. 'What's she want?'

My mum never came unless she needed something. I

wanted to lock the door and pretend there was no-one home, but I knew she'd find a way in. She always did. Once she even broke a window. I sat back down at the table and fiddled with my books. After a minute, Mum stepped in the door. She looked surprised to see us, almost like she might turn back around and leave.

'I didn't know you'd be home, darl.' Her voice was husky and deep, even deeper than I remembered, and she looked from me to Rachel and back again. 'Day off school?'

I nodded, but I was too mad to answer.

'Who's this then? A friend of yours? Introduce me.'

Mum wasn't even looking at me, she was already glancing around the room, checking out what we had. She looked grey, as though all the colour had leaked from her, as though she had a hole.

'Rachel, this is my mum, Danielle. Mum, this is Rachel.'

Rachel stepped forward a little, but still stayed behind me.

'You been in the wars, Rachel? I know how that feels, my back's been playing up so bad, I been like a cripple. Make us a cup of tea, will ya, love?'

I didn't want to, but I stood up and put the kettle on. Searching around for some clean mugs, I could hear my mum flicking through the CDs while my back was turned. After a minute she got game and started opening the cupboards.

'There's nothing here, Mum.' I didn't bother turning around. 'Nothing worth anything.'

'Just checking the place out, sweetheart—haven't been back for a while.'

'Dad was real pissed off about the money box. Gran gave that to me, you know?' My voice felt cold, like to speak to her I had to make myself dead inside.

'What money box?'

'My piggy bank, the one you took last time you came.' I turned around then. My mum's face looked pinched and sad. She had a sore beside her mouth. Just looking at her I wanted to cry.

'See what I have to put up with?' Mum said to Rachel. 'Rock up to see my daughter and get accused of bloody robbery.'

Rachel glanced from Mum to me, holding her broken arm against her chest. I felt sorry for her to be stuck in the middle, but I didn't know how to explain what was going on.

'You didn't even think I'd be home,' I said to Mum, pouring milk into the mugs, willing her to leave.

Rachel leant towards me then, just slightly. 'Is it okay for me to be here?' she asked.

'Yeah.' I looked across at my mum. 'It's better if it's not just me.' Even to my own ears, my voice sounded bitter.

'Better,' Rachel repeated, straightening up and walking over to the couch. 'Okay.' Sitting down, she picked up a book.

'You doing a school project with Gem?' Mum asked Rachel, though her eyes were on me.

'No, I ...' Rachel glanced from Mum to me. 'I crashed my

car, and Vincent ...' The words seemed to get stuck inside Rachel's mouth. She stopped mid-sentence. 'Vincent ...' She looked down at her lap. 'He ...'

'Sugar?' I asked loudly, trying to distract my mum. It was strange, but I felt a bit protective of Rachel.

'Two please, love, big ones,' Mum said, standing there staring hard at the top of Rachel's head. I could see her weighing things up, trying to figure out what was going on. 'You got a mop of hair there, Rachel. Bet it's hard to keep in line,' she said finally.

I brought the teas over, sitting them on the table.

'Gemma braided it for me, for the funeral. So now it's all crimped.'

'Who died?' Mum's eyes were sharp.

'Frankie.' Rachel pushed the book aside. 'My baby.'

My mum had no answer to that, and the room seemed to fill with silence. I wasn't sure what to do, but I walked around the table so I was closer to Rachel. She looked small and pale, sitting up stiffly on the couch.

'So Mum, where you been? Where you living now?'

'Love, I got a place down the coast a bit. Just a small joint, but near the beach. Been living with this bloke called Jackson.'

'That his car?'

'Yeah, well a mate of his really. Just borrowing it for today.'

She looked sideways and then back at me. 'You should come down and stay. Maybe in a few weeks when we're settled. There's no spare room, but we could rig you up a mattress on the floor in the lounge room.'

I crossed my arms. There was no way I was going to stay with her.

'It's hard, Gem, me with all my back problems, I can't work. And Jackson's got bad credit. Nothing to do with him, it was his fucking ex, you know? Bloody bitch spent up big on his credit card. It was hard even to get a place. The rents are so high nowadays. Vinnie's lucky to have this place, such low rent. Friend of a friend and all. You don't know what it's like for the rest of us.'

My mum's story never changed. It infuriated me to hear it again. I turned away, reaching out and running a hand across the kitten's back. It jumped down from the table to the chair, and then onto the ground, sitting beside Rachel's feet. Rachel lifted her legs out of the way, crossing them in front of her on the couch. The kitten turned then, glanced up at her face and jumped straight into her lap. She lifted her arms up, like she didn't know how to touch it, and it circled a few times and then lay down, purring. Even though I was mad at my mum, watching the cat still made me smile.

'You don't like cats, Rach?' I asked.

'I don't know.' She tucked her hair behind her ear. 'I've never had one.'

Mum put her mug on the table and turned towards the bedrooms.

'Listen, darl, I've just got to get something from Vinnie's

room. He's got something of mine in the cupboard in there. I'm just going to grab it and then I'll be out of your hair.' Her voice was muffled as she closed the door behind her.

'That's your mum?' Rachel said, still holding her arms away from the cat.

'Yep, bet she's in there hunting for Dad's pot. She won't find it. He knows better than to leave it somewhere she'd look.'

Rachel nodded and I wondered if she'd understood a word I said.

'Freckles likes you. Listen to how loud her purring is. Funny how animals always want to smooch people who don't like them.'

I bent down and scratched the kitten behind its ear. It was soothing somehow to hear it purr. Mum came back out, eyes scouring the room.

'Did you find your stuff?' I asked, checking out her clothes to see if she'd snuck anything inside them.

'Nah, must have moved it, hey. Oh well. Have to grab it next time.'

I could feel my face tighten.

'Thanks for the tea, love,' Mum said, and pulled me into a hug. 'Better get moving. Told Jackson I'd be home in the afternoon.'

I pulled away from her, so mad I could scream.

'See you, Rach, catch you next time.'

I watched as my mum slipped some CDs into her pants on her way out, not looking back.

Rachel stared at the closed door as the car revved outside.

'She took the CDs,' she said softly. 'Why did she do that?'

'I was waiting for her to do it,' I answered. 'We were right here watching.'

'Maybe she thought they were hers.'

'She hasn't lived here for years. Nothing is hers.'

I walked back around and sat at the table, shuffling my books. I didn't want to think about my mum. I felt like having a shower and washing her from my skin. It was impossible to concentrate on my assignment, so I sat there just tapping my pencil.

'How old are you, Rach?' I asked finally.

'Twenty-one.' She ran one finger lightly along the kitten's belly. 'Twenty-two in August.'

'That's pretty young to have a baby.'

'Same age as my mum when she had me.' Rachel looked sideways at me, watching my face.

'Danielle was nineteen when she had me.' I don't know why I was talking about my mum—it was my least favourite topic. 'I've heard the kids at school say she didn't even know who my father was. But she did. He sent me a card once for my birthday. Name's Greg. Greg Cook. How's that?'

She looked startled. 'Vincent's not your dad?'

'Didn't he tell you that?' It didn't surprise me. Dad usually keeps things pretty close to his chest.

She shook her head. 'He always calls you his daughter.' The kitten pushed out a paw, pressing it against Rachel's hand, like it was trying to get her to stroke it.

'He's my stepdad but he's like my father. I've lived with him since I was real little.'

'I wish I knew something about my father.' Rachel placed her finger carefully against the kitten's nose. 'Anything.' The kitten squinted its blue eyes, looking up at her face like it was listening.

'He must have been dark or something, hey? Must have been something?'

'I guess so. I never looked much like my mum, that's for sure.' She glanced at me across the room. 'In Sydney no-one really commented on the way I look. You know? We moved quite a bit, and no-one ever asked me where I was from, but here everyone does.'

Apart from that time at my Gran's, I'd never really heard Rachel talk.

'It's a small town, I guess,' I said, thinking about her words.

'All the nurses asked me in the hospital here. In Sydney, I think everyone just assumed I belonged somewhere else. Like the Indian kids thought I was half Thai and the Asians thought I was part Maori maybe. No-one said anything, but it was like they all believed I belonged to another tribe.'

'Who'd you hang out with, then?'

'There were always a couple of kids like me, part something but with white mums. One girl, Lucy, her mum was Australian and her dad was Chinese, but from Singapore. We hung out for a while.' She ran her hand across the kitten's back just the once, and it started to purr again. 'But when my mum got really sick I only went to school on and off, so I didn't see her much after that. She used to call me but I never knew what to say. Lucy was going to parties, getting together with boys, wondering if she should cut her hair. But my mum was dying, and I had nothing to say about any of that stuff.'

The kitten purred loudly, its body floppy and unresisting on Rachel's lap. I looked down, swirling my pencil on the page. Rachel talked in a different way from other people. I liked it. She was talking about the things I thought in my head, things I couldn't find a way to share.

'Sometimes I feel like everyone is speaking a different language to me, you know?' My words came out soft. 'Like when I speak, what I'm saying is not translating in the other person's head. Like what I'm saying isn't what the other person is hearing.'

'A miscommunication.'

'You know?'

Rachel nodded. 'It makes you feel like an alien but no-one seems to notice.'

I sucked on the end of my pencil, thinking. 'Why doesn't anyone ever talk about it?'

'There's so many things no-one talks about. When my mum was dying no-one at school would even mention her name,' Rachel said. 'It felt so strange because my mum was such a big part of my world, she was all I had, and then she was gone, and her name was lost too.' She stood up, edging the kitten from her lap. 'But what was even weirder was no-one would talk about their mum either. As though "mum" in general became an unspeakable word.'

'What was your mum's name again, Rach?'

'Sarah.' Rachel's voice was quavery.

'That's a nice name, Sarah. I bet she was a nice person.'

Rachel smiled then, heading towards the door.

'She was just a person, you know? But she was my mother.' Holding onto the door frame, she seemed to be steadying herself. 'Did you ever make those boats?' she asked, staring out into the garden. 'At the river?'

'With Dad?'

She nodded.

'Yeah, me and Dad and Uncle Mick.'

'I'm going to look for some leaves,' she said, and she stepped from the doorway into the garden.

I stood up for a moment and watched her wander around outside. She seemed to look at everything as though she'd never seen it before. Almost like this was her first day in the world, like she'd just been born. She paused beside the rubbish pile. Broken pipes all mixed up with cardboard and pieces of random-shaped metal. That pile of stuff was so familiar to me I didn't notice it anymore, even though it lay in a jumble in the middle of the lawn, the grass growing up around it.

Rachel stepped past the pile, out towards the trees, collecting stuff—leaves and sticks and things. I couldn't quite see what she was doing with them, but she looked alright. I sat down again and kept an eye on her while I tried to do my assignment, but my mind kept drifting to what she'd said. I wondered about all the other unspeakable words. About how to reclaim them. And every now and again I whispered 'Sarah' under my breath, though I knew no-one could hear me.

After the funeral something seemed different between Dad and Rachel. She seemed a little bit more normal and Dad had that look about him. That look he'd get when he was about to fall hard. I'd catch him watching her when she wasn't looking, his face all intense. When Rachel was in the room, she was all he saw. It made me uncomfortable at first, cause I didn't know if Rachel noticed it, and I worried that he'd attached himself to a lost cause.

Rachel is harder to read than Dad.

I watched her carefully for a few days, trying to figure out what was going on. I'd get home from school before Dad, and she'd have the leaf boats she'd made lined up on all the windowsills. She'd started making other stuff too. Hanging-type things out of leaves and sticks, weird but kind of pretty. When I got home, she'd be pleased to see me. Her and the kitten waiting in the doorway. I'd never had that, someone to come home to. On the Wednesday she even had a snack ready, cheese and crackers. Must have been hard for her to chop the cheese one-handed, cause it was all chunky. She'd poured me a juice. I sipped my drink and watched her, and then I found myself talking, telling her about Dave, and how he ignored me at school now. I told her about the fight we'd had at Mel's, what Dave said about my mum.

Rachel listened, not saying much, and then I told her about Mel, about how she'd slept with Matt. I told her how he'd rushed out afterwards, how keen he'd been to go.

'Did she like it?' Rachel asked me, not looking away from my eyes.

'She said it hurt,' I replied. 'Did it hurt you the first time?'

I felt a bit funny talking about sex with Rachel, but who else did I have to ask?

'Yeah.' She ate one of the crackers she'd made for me, slowly, thinking.

'It shouldn't hurt,' she said finally. 'If it hurts, I think there's something wrong.'

'I wanted to run when Dave kissed me. Not straight away, but after a little while. I felt bad. I didn't want to hurt his feelings.'

She watched me, still chewing slowly. 'But what about you?' She poured herself a juice. 'What about your feelings?'

'What do you mean?' No-one ever asked me how I felt.

'I just think it matters,' she answered. 'More than his feelings.'

'Right,' I said, but I was pretty baffled.

'I didn't know that before Frankie died, but I do now.' Rachel looked straight at me, clear-eyed, as though she wasn't some broken-up girl on the edge of trouble. I couldn't hold her gaze. It was hard to think about how I felt. She stood and picked up an apple from the fruit bowl, cutting it jerkily in half, and then in half again. I knew she was cutting it for me.

'What did you get up to today, Rach? What did you make?' I looked around at the scattered leaves.

'Hanging stuff. I hung it out there on the tree.' She pointed out the window and I ducked to catch a look.

'Pretty. They look a bit like mobiles, hey? Sometimes you see those ones made of shells. They look a bit like that.'

'Yeah?'

'Yeah.'

'It's just, all the leaves go brown. They curl up and float away.'

'Leaves don't last long off the trees, Rach.'

'No. They die. Everything does.'

I looked out the window.

'How was school?' she asked, watching my face.

'Good.' I picked up a piece of apple. 'You eat some too. Did you have lunch yet?'

She took an apple quarter.

'This girl at school was telling me and Mel that we've got to play harder to get with boys. That guys will just shag us and move on to some other girl unless we act like we're not that interested. To keep them keen. It was on *Oprah* or something.' The words seemed to fall out of my mouth. 'That's what they all seem to do. Everyone else. Like everything is always an act. And I just can't help thinking it might all work a bit better if people were straight about it.'

Rachel was silent a minute but I knew she was thinking.

'There's some kind of death in not being yourself,' she said finally, her voice strangely solid. 'Maybe ...'

She arranged the leaves on the table, unable to keep her hand still.

'It's like wearing a shroud.'

'I don't know. What's a shroud?'

'The sheet you wrap up the dead in.'

I chewed my piece of apple slowly.

'If everyone is playing a part then how do any of us know what the other person really feels? I mean, that's what I think.' I picked up one of the crackers she'd made me. 'It seems pretty simple. Why do they have things like that on *Oprah* anyway?'

Rachel shook her head and the kitten jumped onto the table, strolling towards her. Freckles liked being near Rachel. She ignored the kitten mostly, but sometimes I saw her give it a pat. Carefully, as though she wasn't sure what it might do. When Rachel sat down on the couch, Freckles always clambered right into her lap. Then Rachel would give it a stroke and I'd hear the kitten purring.

'She just loves you, doesn't she? She thinks you're her mother,' I said, smiling. There was something about Freckles that always made me grin.

Rachel scratched the kitten behind the ears but she looked suddenly sad.

'That's what you don't like, isn't it?' I reached out and ran my hand along the kitten's fluffy tail. 'I didn't get it before.

I thought it was a bit odd that you wouldn't like a kitten. I mean, everyone loves baby things.'

'It's not Frankie, that's all,' Rachel said, looking at the cat.

I reached out a hand to touch Rachel's hair.

'I'll get my hairbrush, hey?' It was the only thing I knew to do.

Standing behind her, I drew the hairbrush softly through, ends first, pulling out the knots. Rachel closed her eyes for a long moment.

'It's something, this house,' she said. 'There's something in it.'

It was a strange thing to say, I guess, but standing in the kitchen, just for that minute, I think I knew what she meant. I wanted to ask her about my dad. About whether or not she liked him the way he liked her. But I didn't. After I finished brushing her hair, I poured us both another glass of juice and sat down opposite her at the table. She picked at the fraying edge of the plaster. It looked a bit the worse for wear.

'Let me colour it, Rach.' I stretched over and grabbed my pencil case out of my bag. 'What do you want me to draw?'

She looked up at me and then back down at the plaster. She seemed surprised it was even there.

'Whatever comes into your mind.'

Pulling off the sling, she stretched her plaster across the table towards me. I closed my eyes and thought. I was blank, nothing in my head at all, and then I remembered the butter-flies I'd seen last summer. A pair of them had floated above me when I walked up the driveway after school. It surprised me that they came to mind, sitting there in the cold kitchen with Rachel. I'd stood still in the driveway and watched them flutter. They'd stuck by me a while, as though they knew I was alone. Blue and black, they knocked against each other in the air, twisting and flapping. They looked beautiful and free and I'd wished for a minute I was like that, and then I'd remembered they only live for three days, and everything had seemed so brief. Like bursts of brightness in the dark.

When I opened my eyes, she was watching me.

'What did you see, Gem?

'I saw butterflies. Should I draw them?'

'Yeah.' She tried to unzip my pencil case. 'What colours?'

I grabbed a lead pencil. 'I want to do the outlines first. I want them to be beautiful. Not tacky, okay?'

Sitting on the table, the kitten watched me draw the outline of the butterflies, reaching out her paw to touch the pencil now and again, softly though, not playing. Rachel laid her head on the table and closed her eyes. I drew and drew, and after a while she started humming her lost baby song. When the pencil drawing was finished I started filling in the colour, and after a bit it looked like I wanted it to. Two butterflies, blue and black, twirling together, just how I'd seen them in my mind.

Dad came home and when he walked in she looked up and I saw something there in her face, some kind of longing. Dad stood in the kitchen, all covered in bits of grass from the brush-cutting, and looked at my picture.

'That's real nice, Gem.' His voice was gentle. 'Real pretty.'

After he got the fire lit, he jumped in the shower and I started on dinner. Rachel helped me cut up the veggies, but she was quiet again, I wasn't sure why. When Dad came out all fresh and clean, he took the knife from her and started chopping. She stood there beside us in the kitchen looking stranded, the kitten circling her feet.

'Rach, you right?' Dad said, looking sideways at her while he chucked the veggies into the pan.

She shook her head, looking down, tracing the lines of my drawing on her plaster. I could see the tears gathering in the corners of her eyes.

My dad did something then I'd never seen him do with a woman. He stepped towards her, wiping his hands on his pants. Then he reached out, pushed Rachel's hair back from her face and kissed her. Gentle but sure, straight on the mouth. I didn't know where to look. I just tried to pretend I couldn't see, but they were right there beside me. She clung to him, crying, her body shaking, and he just kept kissing her, swallowing up her tears. It shocked me. I've been around Dad and his women for years. I've walked in on him having sex a couple of times but it was nothing like that. Watching him kiss her made me so lonely I wanted to cry too.

I scooped up the kitten and went to my room, leaving dinner half made. Curling up on my bed I held Freckles tight, but she wriggled out of my grasp and then it was just me lying there, and so I hugged myself instead.

Dad called me back out when dinner was ready. I was hungry by then and the kitten had been scratching at my door to get out. I fed it some cat food and sat down at the table. I didn't want to look at Dad and Rachel, but I did. Rachel's face was all mottled from crying. She picked at her food, not seeing it, but Dad just ate as though nothing had happened.

'How was school, Gem?' Dad asked me.

'Alright,' I said, taking a bite.

'Just alright?'

'What do you expect? It's school, Dad.' It was what I always said, but that night it sounded hard.

'You okay?'

I didn't know what to answer then. He looked at me as though he was daring me to speak, but what could I say? He was my dad, not my boyfriend. He could kiss who he wanted.

'Dave kissed me the other night. When I was at Mel's.' The words slipped out of my mouth.

'Dave?' He put down his fork. 'Marie's Dave?'

'Yep.'

'What the fuck was he doing there?'

Rachel stood up then and walked across to the couch, pulling the doona up over herself.

'Him and another guy just rocked up, out of the blue.'

'And you let him kiss you?'

'Yeah.'

'Did you let him do more than that?'

I thought of Dave's hand rubbing up and down my back, but I shook my head.

Dad pulled his hands through his hair and glanced over at Rachel. 'Gem.' I could see him struggling with what words to say. 'Sixteen's so young. Too young.'

'Rachel's only twenty-one, Dad, that's just five years older than me. Five years isn't much really.'

'Twenty-one? Did she tell you that?' He looked relieved and I realised he'd thought she might be younger.

'I asked her the other day.'

Dad looked across at Rachel, but she was facing the other way. Staring out the window. The kitten wandered over and jumped into her lap but Rachel didn't stir. Freckles settled down on top of the doona. Dad patted his pockets then, looking for his ciggies.

'Why didn't you ask her how old she was if you didn't know?'

Dad shook his head, as though it wasn't a question worth answering.

'I'm not saying it's right, Gem, me and her,' he said finally, his voice low and strained. 'It just is. Don't ask me to explain it, cause I can't.'

'Are you in love with her?' I hissed back. It seemed a stupid question once I'd said it, but this time he answered me.

'I don't know, Gem. I don't know shit about love.'

'What if I loved Dave?' I felt my voice go all wobbly then, like I might cry.

'Do you?' Dad's neck was red, like he was trying to keep his cool. He stood up, getting out a cigarette.

'I don't know either.' It made me mad that he could never tell me how things worked. 'I'm only sixteen—if you don't know shit, how could I know anything about anything? Right?'

'Gem.' He got out his lighter and put the ciggie in his mouth. 'Maybe I been a shithouse dad.' Looking across at Rachel stretched out on the couch he paused, and when he finally spoke his voice was quiet. 'You've seen lots of fucked-up stuff you shouldn't have, especially when your mum was still around. But you always did good.' He looked back at me. 'I trust you.' I could see it was hard for him to say. 'Just don't mess with shit like that cause you're mad at me, right?'

'I wouldn't do that.' I wished I'd never told him about the kiss. My face was hot.

'Dave's not a bad kid but you want to choose someone who treats you real good, okay?'

I nodded, thinking of Mel and Matt, and wondering if it would ever be any different from that. I must have looked sad, cause he stood up and wrapped an arm around my head. Squeezing me softly in a kind of headlock, the way he always used to.

'You look tired, Gem. You go to bed and I'll clear up.' He headed outside to smoke his ciggie in the cold, and I went to bed and thought my secret thoughts.

Something woke me in the night. I sat straight up, startled, and then I heard it. It was a kind of low moaning. I was frightened at first, it sounded so raw, like something in pain. A dying kind of pain. I listened, my heart pounding, wondering what to do, and then slowly it dawned on me. It was her, Rachel. Dad and her. It was that kind of noise.

I tried to block out the sound but I couldn't. I lay there listening, wondering what they could be doing that made Rachel make that sound. I'd never heard anything like it. My body was tight, lying there, and I tried covering my ears, but whenever I took my hands away the noise was there and I started thinking maybe it was in my head. Maybe it'd been burnt into my brain.

When I got up for school Dad was gone and Rachel was still asleep. I felt strange, like my mind was in a fog.

At school I kept thinking of that noise and at lunchtime I went straight to Dave's hang and sat down beside him, among all the boys. He looked surprised to see me and for a second I thought he might get up and move, but then he put an arm around me and kissed me on the cheek, just like that. He grinned and I smiled back, and when he kissed me properly I didn't care that all the kids could see. I thought of that sound, that deep low sound, hoping that one day it would come from inside me.

21

I took Rachel down to the footy training on Thursday arvo. I should have known it was a bad idea, but she'd been home all day by herself and I could see she felt a bit cooped up. Gem came down too, the three of us bouncing around in the truck. Rachel looked freaked when she saw all the people—the kids roaming around in packs, all the mums with their coffee flasks and fold-out chairs. When I turned off the engine I thought maybe she'd want to stay in the truck but she didn't.

Walking across to where my boys train I could see the women watching us, lifting their sunnies to get a better look. No-one called out like they would usually. Rachel stared around, not smiling. When the kids saw me coming they ran over, kicking the ball my way. I caught it and the boys laughed, mouths flashing white gappy teeth.

Hanging out with these young fellas always soothed me somehow. They didn't ask any tricky questions. They just had fun and mucked about, and every now and then I had to bring them into line. It never seemed like hard work. There was a year when footy training was the only thing that kept me sane. Gem was little, missing her mum, and those Thursday arvos were all we had to look forward to.

Gem headed off to find her friends and Rachel sat and

watched on the sideline, head propped on her knees. I got the boys to do a couple of laps of the oval as a warm-up and then started them in pairs practising their passes. I didn't see Marie come up behind because I was running around with the boys. I turned around when I heard her voice and Marie and a fella were standing on the corner, beers in hand, watching the training. I'd seen the bloke before, drinking at the sports club. Didn't think much of him.

Marie was standing there with him, laughing that laugh of hers, a sudden spurting sound. Too loud. I knew she was trying to piss me off, trying to get under my skin, and it was working. I didn't want to be with Marie anymore, not since I'd hooked up with Rachel, but the sight of her laughing with another bloke made my skin tighten around my bones. I guess I was jealous. I felt like I was puckering inside. They leant towards each other and I turned back to the boys, willing myself to block out the sound of her. I glanced over at Rach and saw she was watching them too.

If I'd let myself think about Marie, I'd have known she'd make a scene. She'd always turned nasty at the slightest thing. I couldn't see her caring one way or another if Rachel got trampled. There was nothing I could do without setting her off, so I just kept on yelling out to the kids, swapping them around and keeping them from going silly.

I didn't like having Marie and the bloke standing there at my back. It was like knowing someone was about to hit you from behind but not being able to turn around and face them. I was waiting for something ugly to come out of Marie's mouth, and while I ran, my brain was stretching itself out to listen for her words. I thought I heard her say my name and maybe the word 'slag', and that was pretty much all I could take. I looked across at Rachel. Her eyes were wide and she was scrambling to get up off the ground. Even though she wasn't that close to them, I guess she'd heard it too.

'Hang on a second, boys,' I called out, jogging backwards. 'I just got to sort something.'

Rachel stood at the sideline watching me, but I headed straight for Marie and the fella. They smirked when I ran up and I felt like throwing a punch.

'What's the go, Marie?'

'Nothing, Vinnie. Not a fucking thing.' She smiled in a twisted sort of way, and I knew she'd been drinking for a while already. They probably both had.

'You got a problem, mate?' The man looked like he was just in it for the entertainment. 'We're not in your road. Don't let us stop you.'

I looked from him to Marie, the blood surging to my head.

'Look, say what you like about me, what the fuck do I care? But you don't know nothing about Rachel, so just back off, right? Go stand somewhere else.'

'Free world, mate, we can stand where we like.' I could see the bloke was up for a fight, but I didn't want it there, in front of Rachel and all the kids.

'You are the biggest piece of shit, Vinnie,' Marie hissed

at me. 'Fucking lying bastard. All that time, and you had a fucking baby with her.'

If Marie was a man, I'd say she would have hit me. Her teeth were clenched so hard her lips were white. The boys had stopped running and were heading over to check out what was going on. I looked around for Rachel. She was standing uneasily further down the sideline like she didn't know where she should be. I wondered where Gemma was. I couldn't see her.

'Marie.' I tried to look at Marie straight, tried to ignore the bloke beside her. 'That's a fucking bullshit story. It wasn't my baby. I'd never met her before. She's not even from around here.' I could feel the mums all craning their necks to get a better view. 'If you want to talk about this shit, then fine. But not here. You tell me when and I'll come round. But I've got the boys here, and Gem ...'

'And her. The fucking slut herself.'

'I told you, you don't know shit, so just keep your fucking insults to yourself, right?'

'You telling me to shut up?'

Marie's hand bounced off my face before I even blinked. The slap stung, the sharp noise of it seeming to ring out across the field.

'You want to grab her, mate?' I said to the bloke, rubbing my cheek and stepping back. I'd never been slapped by a woman before. The whole conversation seemed pointless. 'She's all yours now. You deal with it.'

A couple of men from the club had trotted over, come to intercede. I had nothing to say. It was a lose–lose situation. I just shook my head at them and jogged back towards the boys.

'You're an arsehole, Vinnie,' Marie was yelling behind me, but I didn't turn around. 'A bloody lying mongrel.'

I stopped and stood there waiting, waiting to see if she'd keep going. Throw all her hurts at my back, all the shit she'd saved in a lifetime of dismal fuck-ups. There were no doors to slam and in a bit it was quiet. I knew Marie had moved away, probably gone to get more booze. Abuse me again to her mates inside the clubhouse. The kids all stood around, looking at the ground.

'Come on, fellas, you want to have a practice game?'

'You okay, Vince?' Josh said, leaning down to pick up a ball. He was the liveliest of the bunch but even he looked subdued.

'No sweat,' I answered, but I was looking across the field for Rachel. I couldn't see her. She'd disappeared. I panicked then and started scanning the nearby fields. It had gotten dark. I couldn't see anything that wasn't lit up under the bright lights. 'Just hang on, boys.'

Jogging over to the sideline where she'd been standing, I looked into the darkness at the edge, trying to make her out. Turning back I scanned the crowd of people, all the mums and blokes that I'd been hanging around with since way back. They all seemed like strangers to me. Then I saw Rachel over at the canteen with Gemma. Gem was watching out for me, trying to catch my eye. She waved and I waved back, my heart banging away in my chest. I hadn't been that scared since I'd lost Gem in the supermarket when she was little. I didn't know where I thought Rachel would go, but the weight of that darkness seemed to hang about my shoulders as I ran back to the boys.

'You ready?' I said. 'Divide into two teams, come on.'

'Who's that girl, Vince?' Josh asked me, bright-eyed again. 'The dark girl?'

'She your new girlfriend?' Lochie piped up, still not looking at my face.

'She's ...' I didn't know how to describe Rachel to a bunch of ten-year-old boys. 'She's a friend. Her name's Rachel. You ready to play? It's nearly time to go home.'

The boys shuffled about arranging their teams, and then we played till they'd forgotten Rachel and Marie and the slap. Their hair stuck to their foreheads, damp on red faces. The mums came and stood on the edge of the field, stamping their feet, the cold making them shiver.

'Come on, Vinnie. It's time to wrap up,' one of the mums called out, pointing to her watch and smiling. 'We're freezing our arses off waiting here.'

'Game over,' I called out to the boys. 'Good work, all of you. See you Saturday morning, right?'

The kids wandered off and I picked up the balls and headed to the truck. I knew the girls would be waiting, Rachel and Gemma, and I wondered what they'd say.

'She was your girlfriend.' Rachel hadn't said anything the whole way home, and then just as we rounded the last corner she spoke. 'You had a girlfriend.'

'She was here that day you first came back, remember?' I pulled into the driveway and turned off the ignition.

'There was a woman ... I don't remember really.'

Gemma opened her door and jumped out, quiet, not looking our way.

'It's alright, Rach. It's just Marie. She's fucked up, a bit.'

'I didn't know you had a girlfriend.'

'It doesn't matter now.' I had both hands on the wheel. I wanted to reach out and touch her, put my hand on her leg, but I didn't.

'Because of me?' Her voice was small and the cold air seemed to suck it out the door away from me.

'She was hard work, Marie.'

Rachel looked across at me in the darkness. I could feel her studying my face. I didn't know what else to say, and after a minute she got out of the truck and closed the door behind her. I watched her go inside—she didn't turn back to look at me—and I felt her slipping through my fingers. I couldn't place Rachel in all I'd seen before. I couldn't predict her response to anything. Steam rose from the bonnet of the truck and I sat there watching it, wondering why everything always seemed to go wrong.

22

I thought Rachel might make me suffer, like Marie would have. After the scene down at the footy field, I thought she might hold herself away. But she came looking for me in the darkness that night, like she had all the nights before, and I was glad cause I didn't feel like being punished for nothing.

When we'd finished, she rested her plastered arm on my chest and I looked down at those butterflies, fluttering in the moonlight. Lying there, I didn't feel like sleeping. If I'd been alone I would've had a joint, but I didn't much fancy rolling one up in front of Rachel.

'Gemma used to draw so much when she was little,' I heard myself say. 'Endless pictures, she was always doing them. And then out of nowhere she just stopped.' It'd been niggling at me, that fact, but I'd never let myself say it out loud.

'Why'd she stop?' Rachel's voice was soft, lulled. Her cheek rested against my shoulder.

'I don't know.' I'd never really thought about why. 'I try not to put too much importance onto stuff like that. I try not to read too much into it.'

Rachel looked up at my face but she didn't speak. She slid her fingers along my ribs. I looked down at the butterflies. In the half dark you couldn't see they were just drawn on in

Texta, drawn on by a kid.

My girl, my Gem. Sixteen—not a kid much longer.

In summer round here the butterflies swarm all over, flapping into the windscreen of the truck like love-struck kamikazes. It always seems such a shame. I had to slow down once on the way to work cause there'd been a storm the night before and the dirt road was full of puddles. Came round a corner and there must have been twenty of them at least, sipping from the water. They swooped up in front of me, up into the air, and I was glad I hadn't raced the truck straight over them. Even to me they seemed kind of sacred. I don't believe in that shit, but there were so many butterflies all at once. Sacred in what way, I never did figure.

'Tell me a story,' Rachel said, lying there beside me, looking down the covers.

'What kind of story?' I nudged my foot against hers beneath the doona.

'A story about you.'

'I don't have stories.'

'Tell me a story about having five older siblings.' She rubbed her cheek against my chest. 'Tell me something.'

I thought for a little while, wondering what to say. I wasn't good at telling stories.

'My brothers and me were pretty wild as kids, gave my mum hell.'

The only thing I could think of was the time the creeks flooded.

'Once, after it'd been raining for days, we snuck off with some old body boards to ride the creeks.'

Rachel glanced up at my face. She looked sleepy, but I was wide awake.

'Mum wouldn't let us do that stuff, but we didn't tell her anything. Anyway, the water was flowing real fast and we went tearing down these rapids. Crossing through all these people's farms. When you do that for a while it makes your body turn to jelly, gives you a real high. We were pretty young still, I don't think I was in high school yet.'

'Sounds fun.' Rachel closed her eyes and then opened them again, real slow. 'Like a ride, a fun-park ride, but better.'

'Yeah, well that's what we thought, right.' I rubbed my hand across my nose. 'But we came to this causeway thing. You know, a little bridge with two big concrete pipes underneath?'

'Mmm.'

'Well, we all knew they were dangerous. If you get sucked into them you can get lodged and drown. So we tried to go round—scramble back up the bank and go round. Me and Mick got out of the way, but Gary couldn't get a foothold. We were dragging ourselves out, thinking he was behind us but when we turned around he was inside. He'd gone feet first into this concrete tunnel. The board was wedged somewhere in there. He was stuck. So me and Mick jumped back in to try and pull him out, but the force of the flood was so strong we couldn't shift him.'

'Which one's the oldest?'

'Brother?'

'Yeah.'

'Gary. He's the oldest. He's a good few years older than me. Fuck, he's middle-aged nowadays.'

She rubbed the curve of her foot against mine beneath the covers.

'We knew we needed to go get help, so Mick takes off looking for someone, and I stand there in the water holding Gary's head up. He was getting tired and the water was rising. It felt like forever that we waited. Both of us freezing cold and shit-scared. Gary was white. I'd never seen him helpless like that, he was always my big brother, tough as nails. Anyway, eventually Mick came back with a farmer on a tractor and they towed Gary out. He'd lost his shorts, sucked off in the water, so when we got him up on the bank he was stark naked and so weak he couldn't stand.'

'What did your mum say?'

'She was fucking pissed off.' I pulled my fingers through Rachel's hair. It was so long and dark.

'When we were in there, I could feel the water inching up my body, slowly so at first I didn't notice. In the end I was holding his head pushed right back to keep it out of the flow. It must have caned his neck. My arms were aching, and he was looking at me with this fierce, helpless look, willing me to hold on. I knew I held his life in my hands. It was so fucking intense.'

'Where is he now—Gary?'

'Gazza? He lives in town. Electrician. He's married, few kids, bit older than Gem, they are.'

'You ever see him?'

'Sometimes, you know. At family things. He throws a bit of work my way when he can.'

'Tell me another story,' Rachel said, eyes closing. 'Tell me about your sisters.'

I wasn't used to talking about my family. My mates knew all my stories, there wasn't anyone to tell.

'I bet Mum already told you about Hannah, right?'

'She said you looked after her when your dad died.'

'She say she was schizophrenic?'

'Yeah, I think.'

'Well, no-one really knows. The doctors could never decide on a diagnosis.'

'Tell me a story about her.'

I didn't want to think about Hannah, all the shit that went down when we were young. Back when the world seemed full of promise, when we both believed we'd get out of this town.

'She was pretty crazy sometimes, but then other days she seemed fine.'

I could feel Rachel's body sink against mine. Lying there, I listened to her breaths sliding into sleep and tried not to think about Hannah. The way she smiled when she was right on the edge and it was impossible to tell if she was going to go over.

We had all watched Hannah, looking for the signs that

she was going to go again. With five kids keeping tabs, nothing much got missed, so as soon as Hannah started acting secretive we'd know it was on. And she was so pretty, so wild, that all the boys in town were after her. They didn't get how messed up she was, just thought it was a big bunch of fun. A married man left his wife for her once. Walked out on his children, got a flat up the coast. They moved in together. I could hear the women whispering around town—'He left her for that crazy girl, you know the one?' But the man went back to his wife quick smart, tail between his legs, after he'd spent a few weeks with Hannah.

I tried to breathe in time with Rachel's breaths and let myself drift. It was late, past midnight, but I was wired, my heartbeat clanging. The fight with Marie had injected something uncomfortable into my veins, and my mind kept circling round it. Marie had always been like that. She got me fired up, and then left me clamouring for something. I'd always thought it was her I wanted, that she was my fix. But maybe I'd just mistaken that for adrenaline. Maybe it was the eruption in her I craved, because afterwards I got to put all the pieces back together.

I lay there thinking about Marie and Hannah and Rachel and the fight, and after a while I got to wondering what it would be like to be struck by lightning. A sudden flash of light so bright it was blinding, then a black explosion right inside, and I knew, even if I'd never tell a soul, I knew I wanted to find out.

The next day at work the old guy had us building a veggie garden of all things. He'd gotten wood sleepers and a big heap of organic soil. The boys, Steve and Jimmy, were grumbling about wheelbarrowing load after load, but I liked it. Better to grow something new than all the constant maintenance of old, ailing trees.

My dad had one of those true Italiano backyard numbers, ugly as sin, with cement laid around the edges. We had no grassy lawn when we were kids, it was all fucking tomato plants. He grew everything, had us all out there on weekends weeding.

I'd meant to set up a veggie patch at the house years ago, but there was always something else on. I'd been renting the place for years so I had no excuse, but somehow it had always seemed just like a truck stop, not a real home. I guess I'd thought we'd find somewhere better, me and Gem, but we never did. I had a heap of stuff I'd collected from the dump, stuff I could use on the garden when I was ready. Piping for irrigation, poles to make a trellis, even some old railway sleepers. All in a pile in the backyard, growing grass, like some kind of junkyard bonfire waiting to be lit. Time always seemed to slip away, but the garden was still there, floating in my mind.

Driving home I got a phone call from Scott. He wasn't calling from work, I could tell, cause he called me 'mate'.

'Just thought I'd check in, see how Rachel's going,' he said.

'She seems alright.' I was thinking of her all curled up on

the couch with the kitten. The day before, I'd come home and she'd stuck red leaves all down the trunk of a tree, like the scales of a snake. A winding red river. She'd used the dripping sap. It was bizarre, but kind of beautiful too. 'She makes these things out of leaves and sticks and stuff. A bit weird, but keeps her occupied.'

'A budding artist?' Scott said with a laugh. 'In the city, that would be called "Natural Sculpture" or some shit, and it'd probably sell for thousands.'

'I doubt it, mate.' I held the wheel with my knees so I could fish out a cigarette. 'But she likes it.'

'I heard you had a bit of a scrap down at footy training,' he said. 'With Marie and her new bloke.'

'No comment.' I lit up and sucked in the smoke.

'Copped a slap, I heard.' Scott laughed. I was glad he found it entertaining.

'Yeah.' I breathed the smoke out. 'Marie reckoned the baby was mine. Fucking small-town shit. I got no time for that.' I watched the road ahead, angry all over again. 'Everything with Marie was always a cock-up. I guess it was never going to end well.'

'You sleeping with her, Vince?'

I coughed then, a sudden choking. I didn't expect him to ask.

'Who?' I said when my throat cleared.

'Rachel, you sleeping with Rachel?'

I took another drag on my ciggie. I didn't really want to tell him.

'You dirty dog,' Scott said. It'd been years since I'd heard him speak like that. 'Well, be careful. That's all I can say, Vince. Be very careful.'

'She's different from what she seems.' I didn't quite know how to talk about her. 'Different from other people. She seems a bit flaky but she's not really. She just doesn't care about stuff most people care about.' I paused, trying to get my thoughts straight. 'It's like she's got nothing left, so she's got nothing left to hide. She's free.'

'Free?'

I guess I sounded like a wanker.

'It's hard to explain.'

Scott was quiet on the other end of the phone.

'It's just ...' I wasn't sure how else to describe it.

'I think I know what you mean,' Scott said.

'Anyway, she's alright,' I said. 'She's holding up good. Not jumping at shadows anymore.'

'It's not shadows. It's him. The father of the baby. Men like that don't give up. She's right to be scared. I hear he's still in town. You watching out for him?'

'He's not going to find us up on the hill.' My place was a bugger to find, so I wasn't too worried. If he happened upon us in town—I had it covered. I could protect her.

'Vince?' I could feel Scott trying to figure out if he should say more.

'Yeah.'

'Why'd you always choose the hardest road?'

I smiled at that.

'Dunno. It chooses me, maybe.'

I wanted to ask Scott about the lightning then. About whether or not he ever wondered what it would be like. If he thought about it like I did.

'Thanks for calling, mate,' I said instead and then, 'See ya round. Okay?'

And I curved up my mountain imagining that flashing electric strike.

23

I did it. Sex. But I didn't get close to that sound.

I didn't tell Dad, but Dave had asked me and Mel to a party on Friday night. I went to Mel's after school and hung around, then at about six we told Mel's mum that we were staying at Sally's place and headed into town to get picked up.

When Dave pulled up in his mum's car, it was already packed full of boys and there wasn't any room left for us.

'You'll have to double up,' Dave said and he looked sheepish. I wondered if he'd been smoking pot. I peered in the window at the boys—they all looked dodgy. I guess they'd probably been having bongs for hours. One of them opened a door and Mel and I slid in across their laps.

'If we see the cops, you'll have to duck, right?' Dave said. 'We'll all keep watch.'

Perched sideways on a bloke's lap, I was hunched and my head was knocking against the window. I looked across at Mel. Her eyes looked wary and I realised the fella she was sitting on had his hand on her knee.

'I'm Ben, by the way,' the boy beneath me said. He had shaggy blond hair, longish, a bit curly.

'Gemma.' I swayed against him as we took off.

'You're Dave's girlfriend.'

'Yeah.'

He kept his hands to himself. I didn't know any of the boys in the car except Dave. They were all a bit older than me and Mel. Finished school already. I thought I might have seen some of them around town, but I didn't know their names.

No-one spoke much. With all of us packed in, the car sounded laboured, as if it was struggling to carry our weight. I wondered where we were heading. Out of town it seemed.

'Where's the party?' I asked after a little while.

Dave glanced around, just for a second. 'What?'

'Where's the party?'

'At Simmo's place. You know, out past the turnoff to the miniature horse farm?'

'Yeah.' I felt like I had to shout to be heard in the front, I wished I'd never started a conversation. 'I've never been out that way. Is it nice?'

'Simmo's got a keg,' Dave replied and the boys all grinned.

'I think he's setting up the party in his dad's shed,' Ben said to me. 'It's nice out there, right next to the National Park. Not where the shed is, but—it's just out in the cow paddocks.'

'Cool,' I said quietly back. It was weird to talk to some-one you'd never met while sitting on their knee. Our faces were so close together we couldn't look at each other's eyes. I could feel the slight shifting of Ben's legs beneath me, feel it against the back of my thighs. Even though I'd kissed Dave, I'd never pressed my thighs against his legs. It was freaking me out. I was trying not to think about it.

Across the car, Mel was holding the boy's hand on her leg. The two of them weren't looking at each other either, but I think Mel was holding the guy's hand to stop it roaming around. It made me mad there in the back of the car, mad that Mel had to have some bloke feel her up against her will. Mad that we were stuck in the back. Mad that we'd even come. But we were on the road and we couldn't turn back. It felt like being on a runaway train, sitting in the back of that car, and I kept thinking of us speeding along, getting faster and faster, unable to get off.

It was dark outside, nothing to see. I'm sure Ben's legs were getting stiff, cause when we bumped across a pothole he groaned softly.

'This is it, isn't it?' Dave pointed out a driveway, slowing down.

'That's the one,' the boy in the front answered.

We pulled in and drove along a dirt road for a bit. There were cars pulled up on the side of the road so we knew we must be close. Finally, in one of the paddocks off to the side there was a faint glow. Dave stopped the car and we piled out, then he drove off to find a spot to turn the car around so it wasn't blocking up the road. The boys all clambered through a fence and walked towards the light, except for Ben, who shook his leg out, stamping it a bit.

'Dead leg,' he said to me.

'Sorry.' I rubbed my neck. 'It's cold outside, hey?' My neck was hurting from being cramped against the window. Mel came and stood beside me, grabbing hold of my hand.

'There'll be a bonfire on the other side,' Ben said. 'Come on, you want to come in?'

Mel had insisted I wear one of her tops with my jeans. It was blue with little sequins sewn on. It was pretty but not warm enough and I hadn't brought a jacket. Both of us stood there shivering, our bare arms knocking against each other.

'Come on. Come stand next to the fire. Dave'll find you.'

All we could see in the distance was the big corrugated-iron wall of the shed. The muffled sound of voices hung in the air, and the drumbeat of some tinny-sounding music.

'Alright,' I said and we squeezed through the wire.

The party was bigger than I expected. There was a fire out the back of the shed, just like Ben had said, and people hung about in little clusters, hugging their plastic cups. Ben headed over to the keg to get us drinks, and me and Mel wandered towards the bonfire. There were heaps more boys than girls, and I felt a bit silly all dressed up in the fancy top.

Standing close to Mel near the fire I looked around, trying to see if there was anyone I knew. The bonfire stung my eyes, made the mascara I had put on feel all hard and itchy.

'He's not here,' Mel said, peering around too.

'Maybe he'll come later.' I knew she meant Matt. She'd been talking about him all week.

'What if he doesn't come?' Her voice already sounded

whingey.

'I guess you'll just have to make do with me.' I gave her hand a squeeze.

'That guy in the car was rubbing my leg.'

'I know. I saw.'

'Gross.'

I could only nod my head. I wondered where Dave had gotten to. It couldn't take that long to park the car.

Ben came back over with plastic cups of beer. He handed us one each and took a swig from his.

'You girls ever been to a keg party?'

We shook our heads slowly.

'Awesome. They can get pretty wild.'

I could already hear that sound that goes with people getting drunk. I've been in the pub with Dad enough times to know it. It's where all the voices seem to blur together. Everyone is speaking too loud and all in the one tone. It's an ugly sound. Out in the paddock, even the open air didn't swallow it.

'Where's Dave?' I asked Ben, a bit worried.

'He's probably trying to find Terry. To score, I think. Get some pills.' Ben looked around. 'That was the plan.'

'Right.'

Dave hadn't told me the plan. I thought we were going to the party together, but he hadn't told me anything.

We sipped on the beer, watching the fire. Drinking Jim Beam and Coke was hard enough, but drinking beer was near impossible. It wasn't even sweet and it reminded me of Dad. Mel didn't seem to be struggling with it like I was. She downed hers in no time and handed her cup back to Ben.

'What about you, Gemma? You nearly up for another?'

I skolled the last of it and handed the cup over. The front of me was warm from the flames but the rest was still freezing. I turned around to warm up my back.

'Fuck, why didn't we bring jackets?' I asked Mel when Ben had walked away.

'We didn't know the party was going to be in a fucking paddock.'

'We don't know anyone here,' I said, scanning the crowd. 'There's no-one from school.'

'Lucky Ben's looking after us.' Mel grinned sideways at me. 'I think he likes you.' She giggled and I figured the beer had probably gone to her head, like it had to mine.

'He's just doing a favour for Dave,' I said. 'He knows Dave's my boyfriend.'

'Whatever you say.' Mel tossed her hair, fishing her lip balm out of her little bag. 'This is pretty shit,' she said, smearing on some balm and rubbing her lips together. 'No-one said it was going to be outside.'

'I know.'

'I'm going to get shit-faced. The beer is free, right?'

'I guess so.'

'How we going to get home?' Mel asked, twisting her hair around in her fingers.

'I thought Dave would drive us but I spose not,' I said. 'Everyone's going to be completely fucked.'

'By the looks of it, everyone already is.'

When Dave finally rocked up, we'd been drinking for a while. A couple of other guys had hung around for a bit, sizing us up, but they'd moved on when they saw we didn't have much to say. Mel was looking unsteady on her feet and I was feeling slightly queasy. I hadn't had a refill in a while. I was worried Mel was going to do herself in, the rate she was going.

'Where you been?' I asked Dave.

'Just had to sort something.' He was pale, chewing his lips. He looked strange.

'Did you find Terry?' Ben asked.

'Yeah, bumped into him when I was parking the car. There's a load of them out that way, sorting stuff.'

'Cool,' Ben said and I wondered if he wanted to score too.

'You could get some if you want, mate,' Dave said to Ben.

'Nah. I don't mess with that stuff. Fucks with my head too much.'

Dave just nodded, his eyes darting around. The beer had made my head a bit fuzzy but I could still tell Dave was on something. I've never taken drugs like that. I saw what they did to my mum. It isn't the physical stuff, although that's bad enough. It's what the drugs have done to who she is. It's like she has no feelings left. Like she's hollow. I don't ever want to go there.

Dave put an arm around me but he was stiff and jittery and it felt wrong. I shrugged his arm off and moved closer to Mel. Looking at me with hard eyes, Dave stomped away, not saying a word. I stood beside the fire wondering how we could possibly get out of there.

'I need to do a wee,' Mel said, her voice low.

'Yeah, me too.'

'Just head out behind the trees over there.' Ben pointed into the dark a little way. 'There's no dunny out here.'

'Okay.' I nodded at him, taking Mel's hand. We stumbled across the paddock to the trees and squatted in the dirt.

'Why didn't he come, Gem?' Mel's words sounded damp and twisted.

'I dunno.' I shook my head in the darkness.

'I thought he'd be here.'

'Did he say he was coming?'

'He never talks to me anymore.'

'He's a dickhead, Mel.' That's all I could think of to say. Except for sleeping with her that night, Matt had never really given Mel any indication that he was keen.

'I wrote a song about it,' Mel slurred. 'Can I sing it for you?'

I wasn't sure I wanted to hear Mel's song. I'd heard enough about Matt to last me a lifetime. I stood and zipped up my jeans. Mel looked like she was struggling to stand, so I helped her and fumbled to get her pants pulled up.

'Sing it while we walk back,' I said, feeling sorry for her. 'But do your zipper up.'

'Okay.' She stood there fiddling with the zip. Every movement she made seemed in slow motion. 'Okay,' she paused. 'Here goes—don't laugh.'

I nodded and we started walking. Mel sang so soft I could hardly hear. I missed the first lines and then caught something about 'my heart'. I think she was repeating the same bit over and over.

'Fuck, I've forgotten the next part.' She stopped still in the middle of the paddock, thinking.

'The world's a crock of shit, Gem,' she said out of the blue, eyes foggy, words slurred. 'Dave pissed off. Matt's not even here.' She sat down suddenly on the damp grass. 'Mum's probably at the pub flirting with what's-his-face, Doug. He's gross, that guy, and my mum thinks he's hot.'

I stroked her hair back from her face. She looked like someone on their last legs.

'And your dad's sleeping with that girl whose baby died. It's just fucked.'

'What?' I never talked to Mel about Dad and Rachel.

'Mum said she met the father of the baby down the pub and he seemed like a real nice bloke.'

It was cold in the middle of the paddock. I just wanted to go.

'He's still around?'

'Yep. Mum said he cried when he talked about it. She felt sorry for him. Lost his baby, lost his woman. She's still up there with your dad, right?'

'Yeah.' I didn't know what to say about Dad and his women. It never seemed clean. Always some mess too big to sort out. Then I thought of the sound, and even thinking of it made the prickles start up my legs. It seemed wrong that Dad could have something like that while everyone around him was hurting so bad. An angry kind of feeling swelled in my belly.

'Come on, we'll go back and see if someone can give us a lift into town.'

Back at the fire Ben was waiting.

'Another drink?' he said, watching us approach.

I shook my head, and Mel just plopped down on the ground beside me.

'Look, I've got to get her home I reckon. You know anyone who's going into town?'

He looked down at Mel. 'She looks pretty wasted.'

'Yeah.'

'I'll just go ask around for you, okay?'

'That'd be great,' I said, rubbing my arms against the cold.

Mel leant heavily against my legs. She'd stopped talking now, and I stood there looking around for Dave. Everyone seemed pretty plastered. There was one guy who was standing nearby with no shirt on, not even feeling the cold. The tinny music still played on a little stereo, one you can put batteries in. I might have been imagining it, but it sounded like the batteries were going flat. The music was slightly off, a bit woozy. A couple of drunk girls were dancing, and then all of a sudden they kissed. The boys standing around whistled a bit and then one of them walked over and pulled the girls apart. One girl slapped him and he shoved her, and everyone started yelling.

I looked away. I wanted to cover my ears. I know I must have been drunk, but I didn't feel it anymore, just sort of dull-headed and sad instead.

Ben headed over with a set of car keys.

'I'll drive you in. Johnno said I can borrow his rig, bring it back out tomorrow,' he yelled over the sound of the fight. 'Don't worry about that lot, they've just had too much.'

I looked at the car keys dangling from his fingers and wondered how much Ben'd had.

'You right to drive?' I asked, knowing that we needed a lift, knowing that it wouldn't matter.

'Right as rain,' he said with a smile. 'I've been going easy tonight.'

I nodded and, pulling Mel up, we headed for the road.

There was a ring of them sitting round a smaller fire, out near the cars. Some girls mixed in with the guys. Their talking wasn't as blurred and loud as around the bonfire, but they still didn't hear us approach.

Dave sat cuddled up to one of the girls. I'd seen her at school, she was a bit older than me. Bleached blonde hair.

'I'm just taking Gem and Mel home,' Ben said to the circle. 'They've had enough.'

Passing a bong around they barely glanced up at us.

Ben looked from Dave to me and back again.

'Alrighty,' he said, shaking his head. 'We'll head off then.'

Dave lifted his hand in a wave, and the blonde girl leant in and kissed the side of his mouth. She had a beanie pulled down low on her forehead, almost covering her eyes. She didn't look up at us, didn't even seem to see us there.

'Well, have a good one,' Ben said.

The car was a beaten-up looking red Toyota.

'I thought you were Dave's girlfriend,' Ben said as he unlocked the door.

'Me too.' I felt so tired then, standing there with Mel leaning on my shoulder.

'That's fucked. Shit.' Ben climbed in and leant over to unlock the passenger doors. I opened the back door and Mel clambered in then lay down flat.

'I hope she doesn't spew,' Ben said over his shoulder as I slid into the front. 'Johnno'll kill me.'

He reversed, and soon we were driving back the way we'd come.

'So, you girls got somewhere to crash?' he asked, looking sideways at me.

'Mel lives in town.' Turning around, I peered at her in the dark, 'But I don't know if I should try and sneak her in like that. She's pretty drunk. Her mum'll go nuts.' I didn't know if I was quite up for Mel's mum in hysterics. 'We told her we were staying at a friend's.'

'You can stay at mine.' Ben didn't look at me this time. 'I got a caravan out the back of me gran's.'

I thought about it for a split second but I knew what I would say.

'Yeah, okay. That'd be cool.'

I didn't even try to smile.

I knew I didn't have to sleep with Ben. I didn't have to let him lay me on his floral grandma sheets in the musty caravan. I didn't have to let him take off my top and unzip my pants. To kiss me and lean himself into me.

But I just wanted something.

I wanted it so bad.

I was reaching for that sound but I didn't get near it.

Afterwards, I lay in the dark, thinking about what Rachel had said. About it not being right if it hurt. I thought Ben was asleep but he wasn't.

'Why didn't you tell me you'd never done it before?' he said, turning towards me on the bed. I could see Mel's outline on the mattress on the floor and I hoped she was asleep. We'd had to carry her in from the car, so I guessed she was.

'You didn't ask me.'

Ben hadn't asked me anything. Just laid me on the bed, and that was that. At least he'd put on a condom.

'I just,' he propped his head on his hand, looking down at me, 'I just thought you would have, that's all.'

'No.' I shook my head, wishing I was anywhere but there.

I felt grimy, and a little sick.

'You're beautiful, you know,' Ben said, reaching out and tracing the line of my lips with his finger. 'How old are you anyway?'

'Sixteen.'

'Legal age. That's good.' I could feel his smile in the darkness.

'How old are you?' I asked, closing my eyes.

'Twenty-one. Twenty-two in October.'

He was the same age as Rachel.

'My dad's girlfriend is twenty-one.'

'Yeah?' Ben said. 'How old's your dad?'

'Thirty-six, thirty-seven maybe.'

'That's a bit of a gap.'

I nodded, trying not to cry.

'You alright?' Ben asked, and I knew he could sense me holding back tears. 'You want me to make you toast or something?'

I shook my head, but I curled in towards him. Some comfort was better than none.

Ben stroked my arm gently in the dark.

'You'll be right,' he whispered and kissed me on the head. 'It'll all be cool in the morning.'

He slept then, snoring a little, and I cried and thought of that sound.

That sound I wanted, that I couldn't get near.

I dropped Rachel off at Mum's on Saturday morning before the boys' footy. Mum had offered to make her a nice brekky and I didn't want another scene with Marie. Gem was staying down at Melissa's house, so it was just the two of us. I'd had to wake Rachel up and sort of bundle her into the car. She smiled at me sleepily when I left her, cheeks pink, face bright. Something inside me turned a circle, a quick flip, just looking at her.

Down at the oval I could see Marie watching me from the sidelines like she had something to say. I didn't want another tussle with her so I just got on with it, sorting out the boys' shirts, putting them in their positions. She came up anyway.

'Vinnie?'

I turned around. She looked her normal self, tight-faced, but there was something else there.

'What's up?' I asked, picking up a ball and motioning for the boys to do a lap of the oval. 'You want to slap me again?'

She looked down then, like she was ashamed. I'd never seen her do that.

'I just wanted to say. I know it wasn't your baby, Vinnie.'

This came from left field. I looked at her properly. 'Right,' I said, wondering what had gone down.

'Me and Chris got talking to this fella down the pub last night. He started telling us about how his baby had died, how his girlfriend had left him and crashed the car. He was talking about *her*.'

Even though Scott had warned me, I hadn't thought the man would still be in town. I just figured he must have some other place to be. That he must have a job to hold down somewhere else.

'When I heard him talk about it, it seemed so sad.'

'Her baby died, how could that not be sad?'

She looked up at me, her hard eyes back.

'You didn't have to shag her though, did ya?'

I stepped away, looking at the ground. What could I say?

'I mean, fuck, Vinnie, she'd just had a baby with another guy.'

Looking up I could see the colour rising on Marie's neck. I glanced around to see how close the boys were. They were jogging towards me and I signalled for them to do another lap. 'The boys have got to play in a sec. I can't talk now.'

She straightened up then, like she was tucking away the feelings she had, tucking them out of sight. Like she was taking it on the chin.

'You just race on in, don't you. Race on in and try to patch things up.' Her face was steady. I'd never seen her look so sure. 'But it doesn't work that way. You think you're helping that girl, but maybe you're just using her, like you used me. And what you don't see is all the extra pieces you've left out.

From your patch job. Did you even think of her bloke, of what it must be like for him?'

'You don't know shit, Marie.'

How could I explain about Rachel's bruised-up face, the brown fingerprints on her arm? How could I even begin?

'I know a lot more than you think,' she said, as the boys ran back towards us. 'He was a good bloke, her guy. I could tell. He deserves a chance to work it out with her. You should just get out of the way.'

The referee motioned for me to come over. The game was about to begin.

'You just stick to your life and I'll stick to mine,' I said, turning away from her towards the game. I felt like giving her a whack, just like she'd given me.

'Well, fuck you too,' she said, kind of soft for her, and I could hear her walk away across the grass.

The ref blew the whistle and it was all on. I watched the boys, calling out to them when they forgot their positions, but my mind hummed with a dull noise. A sort of echo that hung there, rolling inside my brain, reminding me of all the times I'd felt like hitting Marie, but hadn't.

When I picked Rachel up from Mum's after the game, she had this scrap of paper in her hands. She was folding it over and over, not looking up. Mum was positively chirpy, wiping down the kitchen benches, and I wondered what had gone on. It didn't take Mum long to tell me.

'I placed an ad, Vinnie, like I said I would. And you wouldn't believe it, but we already got a reply.'

'What?' I couldn't follow what Mum was going on about.

'Rachel's grandma, she responded to my ad.'

Rachel unfolded the paper and handed it to me. A name and address in my mum's curly writing, and at the bottom a number.

'I put an ad in the Queensland paper—you know, in the notices section—asking if anyone was interested in the whereabouts of Sarah Jones's daughter. Got a reply in the first week. Word gets around.'

'Who's Sarah Jones?' I still didn't get it.

'My mum,' Rachel said quietly. 'She's found my grandma for me.'

Handing the paper back, I looked at Rachel. 'Did you ask Mum to do that for you?'

Rachel rolled the paper between her fingers, shaking her head.

'No.' She smiled a cautious smile. 'I didn't even know her name. It's Dulcie. Dulcie Jones.'

Mum was beaming, looking real proud of herself.

'It was so easy, Vinnie. I had no idea if it would work, but it did.'

'Mum, did you ever think of asking?' I pushed my hands up through my hair. Mum had always been a bit of a wild-card, taking things further than I thought she would. 'Did you ever think maybe you should find out what Rachel wanted?

Maybe she didn't want to know her grandma's name.'

'You're not going to be a stick in the mud about this, are you?' Mum put her glasses on, looking at me closely. 'Rachel, love, you're happy she rang me, aren't you?'

Rachel looked from Mum to me, and back down at the paper.

'It's okay,' she said, folding the paper up and putting it in her pocket. 'It's okay, Vincent.'

There must have been a place in me that knew then and there that she would leave me. With somewhere else to go, she was free. I didn't like it, my gut felt hollow. But it wasn't my business—Rachel's family—so there didn't seem anything else to say.

'You want a snack before you go, Vinnie?' Mum stood watching me, her hand on the fridge door. 'We already had some cake. Banana. I'm trying out some healthy recipes.'

'Alright,' I said, letting the whole thing go. Mum got out a plate.

Rachel was staring down at the floor. I kept thinking of that scrap of paper folded in her pocket. Such a fragile white slip, and written on it the place she might end up when she'd had enough of me.

'All my leaves go brown, Vincent.' Rachel's voice was muffled, she was talking to the floor. 'I wish I could keep them red.'

'What?' I crouched down in front of her.

'All my leaves, they're so bright and red, but they go brown. They curl up.'

I was listening but I wasn't sure what she was getting at.

'In a few days, it's all gone.'

'What's gone, Rach?' I wanted to lean forward and kiss her, but I was crouching in my mother's kitchen and I knew Mum was listening.

'All the life.'

I knew then that in some way she was talking about Frankie, and all the leaves and sticks and things she collected were about him too.

'You talking about autumn leaves?' Mum said, handing me a piece of cake. 'Vinnie knows how to keep them red, don't you, Vinnie? Just whack them in some metho in a jar. They'll keep.'

Rachel turned to Mum. 'Metho?'

'Yep.'

'Do you have any, Vincent—any metho?' Rachel stood up then, like she was ready to go.

'Under the sink at home, I reckon there's some.' I popped the piece of cake into my mouth and we headed for the door. Mum followed us out and Rachel scrambled into the truck.

'Thanks, Mum.' I brushed my hands on my jeans and Mum waved to Rachel through the window.

'Vinnie?'

I knew that tone of voice. She was going to try to tell me something.

'Everyone needs options, darling.'

I stood there thinking about that.

'That's what her grandma is, an option?' I asked, my hand on the door handle of the truck.

She nodded, leaning forward to give me a hug.

'Just an option, Vinnie. She might never take it up. But she can. See? It makes all the difference.'

'Okay.' It was best not to argue with Mum.

'You got to be more than a knight in shining armour, darling. It's got to work outside that.'

'Okay, Mum, I get it.' I jumped into the truck and Rachel smiled across at me.

'Metho's clear, right?' she asked. 'You have jars? With proper lids?'

'Yep. It's all there, under the sink, at home.'

It amazed me that something so small should brighten her day, and I couldn't help but grin back.

'Thanks, Vincent,' Rachel said, and then she leant over and kissed me on the neck. 'Metho. In a jar.'

I waved to Mum as we pulled out, and I could see her grinning. It would piss me off normally, my mum's interfering, but nothing seemed to matter in that moment except me and Rachel. The dull noise of Marie's words inside my brain lifted.

We had everything we needed.

Metho and a jar.

25

Back at home Rachel went straight to the sink and brought out from underneath a big two-litre bottle of metho, mostly full. She smiled and pulled out the jars.

'Six.' She counted them. 'And they all have lids!'

'Knew they'd come in handy some day.'

'Come find some leaves with me.' She was already half out the door. I followed her past my pile of garden stuff, out towards the trees.

'What are we looking for?' I asked, scanning the ground.

'The bright ones. Yellow or red. Or even green.'

I nodded. The backyard was cut into the side of the mountain, so there was a forest that kind of loomed down over us. Some of the trees had changed colour and were dropping leaves. That's what she was after. We scrounged around a bit, finding a few. She held them fanned out in her good hand, staring down.

I was thinking about my pile of garden stuff, thinking maybe it was time to make a garden. I knew exactly where to put it—I'd picked out the spot ages ago, full sun in the mornings. I left Rachel looking for the leaves, and collected up the bits and pieces I'd need. In a while she wandered back into the house, leaves in hand, and I started digging trenches for the sleepers.

I heard the car pass by on the road, I guess, but I thought nothing of it. When her old man walked up the driveway, all spruced up with a bunch of flowers, I was surprised. I strode towards him, cutting him off, the shovel still clenched in my hands.

'What are you doing here, mate?' I said, and I could feel the muscles in my arms tightening.

'I've come to see Rachel. I've come to talk.' Paul peered past me towards the house. He had on a pair of slacks and a blue ironed shirt. Like someone who worked in an office.

'She doesn't want to see you. I told you at the funeral.'

'Rachel!' he called out. 'I just need to talk to you!'

I looked at the flowers and then glanced over at the window.

'You have to leave, mate. She's not going to come out.'

The curtains shifted slightly and I knew Rachel was hiding there, watching. I felt strong, standing there with my shovel, the sweat dripping down my back. The man glanced from the house back to my face, swapping the flowers from one hand to the other.

'I haven't come to make trouble. I just want to talk to her.' He looked pitiful, dressed up and tired, pleading with me in the hot sun.

'Mate, she doesn't want to see you.' I had no sympathy for him. 'You're not welcome here.' I wasn't backing down. It was a stand-off. I bounced the shovel in my hand, feeling a bit impatient. I could see the man's frustration. He clutched the flowers so hard his knuckles turned white.

'Rachel!' he called out again. 'Please!'

No sound came from inside. The curtains hung unmoving against the window but I knew she was still standing there watching.

'Go home, mate,' I shouted, my voice loud in my own ears. 'Go back to Sydney. Stop hanging around. She doesn't want to fucking see you, and she's not going to change her mind just because you bring a bloody bunch of flowers.'

I could see he wanted to hit me, but I was standing there with the shovel, and he knew he didn't have a hope in hell.

He turned around and I watched him stamp back down the driveway. Swearing, he threw the flowers into the garden, kicking at the shed wall as he went past. I watched him go and then listened for the sound of his car driving off. When I knew he was gone I looked across at the window, and Rachel was there, staring out at me. I lifted up the shovel in a kind of wave, and she held up one of her jars, the leaves swaying inside. Even though I couldn't see it through the glass, I knew her hand was shaking. I smiled at her, wiped the sweat from my forehead with my shirt and then headed back to the garden.

Paul had gone so I wasn't expecting trouble. The shovel cutting into the dirt was loud and I didn't hear anything until a hissing sound rushed behind me, and I turned around in time to cop a whack to the head.

When I came to I was still lying out in the garden beside the shovel, my hands tied tight behind me. It felt like wire, like it was cutting off my circulation. My head throbbed and it took me a few seconds to realise what was going on. That he'd come back.

With my hands tied like that I had trouble even standing up, but when I finally did, I staggered over to the house. I wanted to just check things out, but when I saw what he'd done I stumbled inside.

'What the fuck?' I heard myself shout.

Rachel was sitting on a chair and Paul was cutting off her hair. She turned towards me, eyes smudged-looking, cheeks red.

'Come any closer and I'll kill her,' he said, holding the scissors to her neck. The man's whole body was shaking. He'd been crying. They were my scissors—he must have found them in the drawer.

'I'll kill her.' He pressed the pointed end of my scissors against her skin.

'What the fuck are you doing?' I wasn't thinking straight to ask that. He yanked at Rachel's hair, pulling her head sideways. I saw then that her nose had been bleeding, that her face was swollen, and I backed away against the wall.

'Sit down,' he said. 'On the floor.'

If I'd had my hands free I would have done something, anything, but I slid myself down the wall instead, tasting blood in my mouth from where I'd bitten my cheeks. The throbbing in my head seemed to expand into the room.

'Don't hurt her,' I said through gritted teeth.

'I can't believe she left me and now she's living here with you. It's like I never even happened. It's like Frankie never happened.'

He took the scissors away from her neck and pulled up a chunk of long hair. I watched as he snipped the hair off at the roots and threw it onto the ground. Holding back tears, he wiped at his eyes with the back of his hand.

'I didn't want trouble. I just wanted to talk. To talk to the mother of my baby.' He touched her hair again, looking down at it. 'My dead baby. But you wouldn't listen.' He looked at me, running his fingers along the length of her hair. 'And when I came in here, she tried to get away. She wouldn't even sit down. My baby. It's all over in one night, and I don't even get to talk to her. And here she is playing happy families with you. You in the garden, her in the kitchen. What have I got?'

His eyes were red, filling up.

'I have a right to hear how Frankie died, a right to hear it from her.' Tears spread out along his cheeks. He wiped his nose on his sleeve. 'You slept with her, right? And my baby's been dead how long?'

Rachel was silent, eyes on the floor.

'Three weeks? When did she first fucking sleep with you? The night he died?'

I couldn't look at his face.

'And I'm the fucking bad guy, right?'

A kind of fury swelled inside me. He looked at me then, like he could see it on my face.

'Not the hero now.' His voice was soft, his eyes narrowed on my face. 'You think you're different from me.' He pulled up another clump of hair. 'You're not.'

I wanted to tell him he was wrong but no words came out.

'You helping her? Giving her a hand?' he asked, and Rachel winced as he sliced through the hair. 'That's what I did.'

My hands tugged against the wire, I could feel it cutting into my skin.

'You in love with her?' he asked me, throwing a ball of hair my way. It landed beside me. It was dark and dead-looking on the floor, like an eel beached up on the shore.

'I am,' Paul said, smoothing the half of her hair he'd already cut. His touch was almost tender. 'I'm in love with her.'

Rachel's eyes were glazed over, like she had shut up shop.

'So when did you start hitting her?' My words felt like grit in my mouth. I couldn't believe I was letting him stand there like he was some harmless stranger having a chat.

'It's not that simple, mate.' He pushed his free hand through his hair, showing up the grey at his temples. 'She kept herself from me.' His voice was low. 'It was like she had a secret garden that I could never get inside. And when she got pregnant with Frankie it got worse. She just closed me out.'

'So you hit her.'

'She kept everything from me. Everything that mattered.'

I wanted to spit at him.

He pulled the last pieces of her hair away from her scalp and carefully cut it off. She had only short tufts left, sticking up at strange angles. The hacked hair was all over the floor around us, black and thick, like an oil spill at sea. I glimpsed the shape of the kitten beneath the cupboard, hiding. Animal instincts kicking in. Rachel's jars were lined up on the windowsill, the coloured leaves sinking slowly in the clear liquid. My hands were completely numb. There was a stillness in the room, like the lull before a storm. It was as though time had stopped.

'I could have looked forever and never found this place,' the man said out of nowhere, looking around the room. 'Bit of a dump, isn't it?'

I shrugged, trying to wriggle my fingers back to life.

The man seemed calmer all of a sudden. Clutching the scissors, he had his hands on Rachel's shoulders but he was looking at me.

'Friends of yours at the pub pointed me in the right direction. They knew I needed to see Rachel. Sort stuff out with her.'

'What?' His words hit me hard.

Paul ran his hand over Rachel's head. 'Those bitches at the hospital wouldn't have a bar of me, but at the pub it was different. I mean, there are two sides to every story.'

'What?' I said again.

'They heard me talking,' he said, looking down at Rachel's hair. 'A man and woman. They came and had a beer.'

'Fucking bitch.' It was out of my mouth before I knew it.

Rachel looked across at me, suddenly alert. There was something in her face, something hard and brittle.

'You said he wouldn't find me, Vincent.' Her words were slow and laboured. 'You said he wouldn't get near me.'

I nodded, my head pounding. She was talking straight to me, as though he wasn't even there.

'But I knew he would. He always did.'

I could see it hurt her face to talk.

'How'd you even meet him?' I asked.

'He lived next door to us in the beginning, in this old apartment block. He always used to watch us come and go. When Mum got sick, we relied on him. There was no-one else. We didn't have a car and he drove us around sometimes. To the hospital for the chemo, and all the scans and stuff. It was endless.'

'Seemed an easy way to help out,' he interrupted, watching my face.

'When Mum died I couldn't get it together. I tried to get work and stuff but I couldn't cover the rent.' She kept looking at me. 'Paul let me move in with him.'

I didn't want to hear it.

'I didn't know what it meant. Didn't know what he was after.' Twisting her neck, she faced him. 'You were always watching us, even when we first moved there. I thought it was Mum, Mum that you wanted.' Her voice was ragged.

'I was Sarah's friend. I looked out for her.'

'She trusted you,' Rachel said. 'You promised her you'd

look after me. When she was dying, you promised her that.'

'I did look after you.' The man started crying again. Softly weeping. 'And you ran away with Frankie.'

I felt apart from them somehow, like a glass wall had grown around them and I was watching from outside.

'And now Frankie is dead.' She closed her eyes. I could see her breath move in her chest and then she opened her eyes again.

The man's body was shaking and he brought his face down real close to hers.

'All I ever did was love you.' His voice was low.

I could see Rachel's body tensing, poised to move.

'It's not love,' she whispered back. 'It's not.'

She smacked Paul then with her plastered arm, quick as a whippet—right on the bridge of his nose. She must have caught him by surprise cause he staggered to the side, and she was up and running, straight out the door.

Paul looked at me a second, tears spilling from his eyes, his hand cupping his nose, and then, still clutching the scissors, he bolted after her. Shimmying myself up the wall, numb-armed and clumsy, I followed.

When I got out to the road, Gemma was there, holding a star picket up like a sword, keeping Paul at bay. She had on this tiny slip of a shirt, sparkling with sequins, and I stopped short cause I'd never seen her look like that. Rachel was gone. I guessed she'd taken off down the road.

Star pickets are long heavy stakes of steel, and Gemma's arms were shaking, but she didn't look scared. She looked fucking angry.

'You cut off her hair, you bastard,' she yelled. 'How could you do that?'

I ran around Paul to get to Gem. He didn't try and stop me.

'I called the cops,' she said. 'I could see you through the window.'

'They coming?'

'Yep.' She didn't take her eyes off Paul. 'It was Roger. From footy. He'll be here in a minute.'

Paul stood there squeezing the scissors tight in his hand. Like he was tossing up between launching at us or running. A line of blood dripped from his nose. I could hear the hum of a car in the distance, slowly getting closer, and I hoped Gemma's arms would hold.

The police car pulled round the corner, quietly creeping to a stop. Roger and his mate stepped out.

'You right, Vince?' Roger asked but he was looking at Paul. 'We got backup coming. We weren't sure of the weapon.'

Paul dropped the scissors and ran. The two cops took off after him and Gemma sagged forward, the star picket hitting the ground with a clang.

The back door of the police car opened.

'They've got her already,' Gemma said, and running over she ducked her head in the door. I couldn't hear the words my daughter spoke to Rachel, but she turned to me and said,

'She needs some ice. For her face.'

I nodded but I couldn't move. My feet wouldn't budge.

'Dad, you come and stand with her while I get some.' Gemma ran inside and I stepped towards the open car. My hands were still tied together behind my back. I felt useless standing there with nothing to offer her.

'Frankie died here,' she said, hiding her eyes. 'Right on this spot.' With all her long hair gone she looked even smaller than usual. Childlike. Her face was puffy and swollen, already shadowed on one side from where he'd hit her.

'I know, baby.' My voice cracked and I knelt down there on the road.

'And he's all burnt up now. Turned to ash.' She moved across the seat a little, closer to me.

I couldn't put my arms around her or touch her face, so I lay my head in her lap. Her legs felt bony beneath my ear, bony and cold, even through the jeans. She put her good hand on my head and I closed my eyes, blocking out the light. Murmuring something, almost a song, Rachel moved her fingers through my hair. Beneath my eyelids I felt the liquid pool, my mouth filling up with wet salt.

'It's not love,' she breathed. 'Is it?'

I stayed still beneath her hand, tears easing from my eyes.

26

When I woke up on Saturday morning after the party I felt pretty shit. The walls of the caravan seemed to press in against me, the smell of beer and wood smoke seeping up from my skin. Mel groaned on the mattress on the floor and then she stumbled outside and I heard her throwing up into the garden. A bit woozy, I clambered out of the bed, pulling on my clothes. Ben turned over, squinting at me as though a light was shining right into his eyes.

'Don't rush off.' He pressed his fingers against his eyes. 'I'll give you a lift.'

It felt weird to even look at him. He seemed larger than the night before, a man not a boy.

I nodded but headed for the door. 'I better give her a hand,' I said, not knowing how to act.

Mel was crouched shivering on the edge of the garden, hidden in the foliage. She looked bad, her skin almost green, and I wondered how I could possibly help her.

'Fuck, Gem,' she said, teeth chattering.

'It was too much.' I crouched down and pulled her hair back for her. 'You'll know next time.'

Her whole body convulsed and she spewed again into the garden.

'I am never ever drinking again,' she said but her colour looked brighter.

I sat there for a while beside her and when she seemed strong enough I helped her inside.

Ben had made some Vegemite toast and poured out a couple of glasses of orange juice.

'You'll feel better if you eat something,' he said to Mel. 'Everyone says greasy stuff does the trick, but I reckon Vegemite toast is best.'

There was a small table that came down from the wall, and Mel and I squeezed in around it and ate the toast. Ben made himself a coffee and stood in the doorway drinking it. I could feel him watching me but I couldn't make myself look up.

'My mum's going to kill me. She'll know straight away.'

Mel's hand was shaking.

'I have to take the car back out to Johnno, otherwise we could have just hung here for a bit.' Ben sounded sorry. 'Gotten a DVD or something.'

It hit me suddenly that I just wanted to be home. That homesick feeling thumps you in the belly, a topsy-turvy feeling. I used to get it when I was little and I'd stayed at a friend's house. I'd be going along fine and then—bang—I would just need to go back.

I knew I'd have to sneak in my window so Dad wouldn't see me, but I'd done it before. I felt sort of bruised and sore between my legs, and it was hard to smile back when Ben gave me a shy little sidelong grin. He had dimples on his cheeks that I hadn't noticed last night in the half-dark. Checking the time on his mobile, he patted his pockets.

'Shit, it's twelve o'clock already. We slept in,' he said, finding the keys on the bench. 'Told Johnno I'd be back out by twelve, so I guess we'd better hit the road.'

Ben dropped Mel off first and then headed up the hill to my place. I told him just to drop me out on the road. I could tell he wanted to linger a minute saying goodbye, but by that time I was feeling pretty shaky and just wanted to get away. I kissed him though, cause he asked me to, and watched him drive away. Up on the hill it was cold, and I still didn't have anything but Mel's little top on. I felt parched and crusty, and hoped no-one was home so I could have a long hot bath.

Creeping around the side of the house I heard something. It was a voice I didn't recognise—a man. The sound stopped me short cause we don't have many visitors. After listening for a minute, I slid quietly around the house till I hit the kitchen window. He was cutting Rachel's hair, the man from the funeral, and my dad was on the floor. Staring at the three of them, a heat rose up through my body. I wanted to scream, but I knew I needed to hold it together. I needed to call the cops.

When Dad came bolting out of the house his face looked so broken I thought I might kill the bloke. Inside, the man had strutted around like a king, but outside he seemed to wilt—the reality of what he'd done slowly reaching him. I held that big steel thing up high, ready to crash it down on his head, and I wanted to. I wanted to see that man bleed.

He'd tied my dad up and after the cops had come I ran inside but I couldn't find anything to cut the wire on Dad's wrists. I grabbed some ice for Rachel's face, wrapped it in a tea towel, and then I opened all the drawers, but there was nothing. Running out to the truck, I got the secateurs.

Dad was on his knees with his head on Rachel's lap. She was crooning, softly, and she shrugged away the ice. I tried to cut through the wire, to set Dad's arms loose, but the secateurs weren't strong enough. The blades just got wedged and I had to run off again to try and find some wire cutters. By the time I got Dad free, he could barely stand.

I'd never seen him look that way, like a sailor lost at sea.

When we got back from the police station the house was cold. They'd taken us to the hospital first, done photos and stuff for evidence, and then we'd gone to the station to give our statements. It took hours. They asked me all these questions about Rachel's relationship before Dad. But I didn't know anything. No-one had told me a thing. I could only tell them what I'd seen through the window, and they made me do that again and again.

My body ached and I felt a bit sick. I realised I'd only eaten that Vegemite toast in the morning with Ben and Mel, and then just endless cups of tea at the police station. Home felt all wrong, the place where it had happened, and

I could feel Dad's anger building. At the hospital they'd given Rachel some kind of painkiller and now she sat on the couch, staring at the unlit fire. She was so beautiful. Even all bruised up, hair cut away, she was luminous somehow. Fragile and wounded but still pulsing with life. On the road she'd sprinted past me like a bird that had flown into the glass and swooped off again, dazed and off-centre. Watching her now, I kept thinking of those frozen seconds when she was that bird.

Her hair was still scattered across the lounge room floor, and even though I was sore all over I got the broom and swept it into a corner. I know hair is just dead cells, even when it's growing from your head, but her hair had always seemed special to me. Alive, even when the rest of her had shut down. Whenever I'd brushed it and made all those red strands glow, I'd felt I was doing some kind of magic—bringing her back to the world from wherever she'd gone. I didn't understand how he could do something like that. It was like taking away the last thing she had.

'Rach?' Dad was prowling around the house, unable to keep still. 'Why didn't you tell me about Paul? That he was just the guy who lived next door? Some old perv?'

I didn't know what they were talking about.

'I tried to tell you,' Rachel said, her voice hollow, 'but I can't tell anymore which bits are important and which bits aren't.'

'He wasn't even your boyfriend. You should have told me

what happened after your mum died. That he took advantage of you.'

Rachel looked up at Dad, her face strangely still.

'After Frankie died it didn't matter. Nothing did.'

I wondered whether maybe Dad thought he was the same—that he'd taken advantage of her too.

'It's different, Vincent,' Rachel said, reaching out an arm to him. She must have had the exact same thought as me.

He didn't move to touch her outstretched hand and something inside me cracked.

'Dad, you never even asked her how old she was,' I heard myself shout. Usually when Dad got mad I just let it wash over me but this time I was mad too. 'Did you ever ask her about her boyfriend?'

I knew it wasn't Dad's fault that some other man had beaten her, but in those moments it felt like it could have been. Dad pulled a shaking hand through his hair. There were lines of red on his wrists from where the wire cut in.

'It's not my business what she did before.'

'Maybe.' My voice came out low. 'But you can't blame her for not telling you when you never asked.'

I wanted to run myself a bath to try and soak away the grime of what had happened. Of the keg party and the caravan, of Mel and Ben. Of the father of the baby, and what I'd seen through the window. After today, everything seemed different. I wasn't a dumb kid anymore, on the outskirts of it all. Whatever was happening now, I was inside it.

'You don't know anything about her, do you?' I said. 'If you'd asked her anything, you would have known. She would have answered. And no-one told me *anything*. You should have told me. I didn't even know why she was scared of him. You never told me about it. About him. What if he'd come when you weren't here? I wouldn't have had a clue.'

Dad couldn't meet my eyes. I'd never seen him look so low. The words had spilled from my mouth in an ugly pile— but I couldn't find a way to take them back. He went outside for a ciggie, but he was only gone a minute before he banged in through the door.

'I've just got something to sort,' he said, looking anywhere but at me.

'You going out?' I asked, stressed about what he might do.

Rachel stood up, wobbling on her feet. 'I'm coming,' she said, sounding doped up. 'You're not leaving me here.'

'Me neither.'

Dad looked pissed off but what could he do? We piled into the truck and drove. It was dark and the lights of the truck seemed dim against the night.

'Where are we going?' I was starting to feel scared.

'Down the pub.'

'Why?'

'Marie and that bloke sent him up there to us. They told him where we live.' I'd never heard him sound so angry.

'Dad, anyone could have told him where we live.'

'No-one else would tell a stranger where your house is.

Not in this town. She did it.'

'He's been hanging around talking to everyone, not just Marie. Mel's mum's been talking about him too. Everyone felt sorry for him.'

'Marie sent him up here. I know it.' The truck swerved around a bend.

'Dad, you're acting crazy.'

'He told me it was her,' he growled. 'Her and that guy.'

I could see the muscles in his arms, he was gripping the wheel so hard.

'But Dad, even if it was them, they didn't know he'd cut off her hair,' I said. 'How could they know that?'

'They shouldn't have fucking done it.'

Rachel sat in between us, staring at the road.

'Vincent,' she said softly, 'it doesn't matter.'

'It fucking matters to me.'

I kept thinking of him kneeling at the police car door, arms tied behind his back.

'Can you at least slow down?' I said, trying not to cry. The truck was swinging round the corners. I was afraid all the guilt and fury he felt at himself was going to come out in Marie's direction.

'They just shouldn't have fucking done it.'

At the pub Dad jumped out and headed straight inside. We followed him, me and Rach, stumbling along behind. The pub was bright under the fluoros. It didn't take him long to find them, sitting at the window. I hung back in the doorway, Rachel clinging to my arm.

'Vinnie.' Marie looked surprised.

The bloke she was with stood up to get a beer. 'You want one?' he asked Dad, heading for the bar.

Dad shook his head. Even from the doorway I could see his face going red.

'Do you have any fucking idea what you've done?' he yelled at Marie. 'You don't, do you?'

Marie stood up, like she might try to back away.

'You sent him up to my fucking house.'

The bloke wandered back, holding his beer. 'Settle down, mate, we're just having a few drinks, you know? No need to get wound up.'

'It was his baby, Vinnie. She was still his girlfriend three weeks ago.' Marie was usually so shrill but tonight she just sounded tired.

'You want to see what he did to her?' Moving back towards us, Dad grabbed Rachel's arm, pulling her across the room.

'Look.' His voice was breaking at the edges. 'Look what he did.'

He tilted Rachel's face up to the light and she winced and closed her eyes. I didn't like the way he held her face. I could see his fingers digging into her cheeks.

'Dad, let go. You're hurting her,' I cried out from behind, pulling at his arm. He couldn't hear me. Whatever hold he had on his fury, it was hanging on by a thread.

'You fucking sent him up there to do this,' he said, looking from Marie to the man. 'He cut off her fucking hair.'

Dad's rage seemed to swell, enveloping the whole room.

'What is it with you lot?' he shouted. 'Fill yourselves up with bullshit gossip then sit back and sneer.' He looked around the room, still holding Rachel's face. 'She look like she deserved this?'

I think I knew then why he was so wild. Compared with the rest of them, Rachel was pure. She'd stepped into the fire and come out clean. But with a few words they'd sent Paul our way and he'd dirtied her. They all had.

'Back off, mate,' the man said, holding out an arm. 'It's not our fucking fault.'

Marie was standing still, holding her hand over her mouth, staring at Rachel.

'Dad, let's go home,' I heard myself say. I wanted to get out of there.

'Just look what you've done.' He let go of Rachel and she backed away, clumsily, up against the wall.

'Mate, where were you?' The bloke stepped forward. 'Why didn't you stop him?'

Dad looked down at his hands for a second, like he wasn't sure what they were for.

'I mean, she's your woman, right?' The man's voice was low, goading.

Dad's arms clenched, his hands tightening into fists.

'Look, back off alright, it's got nothing to do with us.' I could see Marie's fella was finished talking but Dad didn't move, just stood there pulsing with some kind of violence.

The man reached out, like he was going to shove Dad on the shoulder and something happened in the room then. It was like a great shadow detonated and Dad just went nuts. He launched himself at the man, laying into him like he wanted him dead. Marie screamed, jumping out of the way, and the man groped around, blindly trying to fend off Dad's fists. There was a thud of flesh hitting the table, blood spattered from somewhere onto the wall, and the pub erupted. Blokes came running from round the bar to pull Dad off, but the fight just seemed to get bigger.

Someone knocked into me and I felt myself go down. There were legs all around, and I tried to push myself out of the fray. I caught sight of Rachel across the room, covering her face with her hands. I wanted to go to her but I couldn't get past all the legs. Through the moving bodies I could see Rachel was shaking, her knees dropping out from beneath her. Someone walked across to help but she waved them away, and then all of a sudden she was up and running out the door.

Pushing myself over, I crawled out from the fight. I could see Dad being held back by a couple of men.

'Dad!' I yelled. 'Dad, she's run out.'

He couldn't hear me over the noise, and in any case his face looked blurred, its edges dissolving. I headed for the door. Outside I could see her silhouette running along the road, awkward and lopsided with her heavy plaster, heading away from the light.

'Rachel!' I screamed, but my voice seemed swallowed up by the night.

A truck roared towards me, its headlights shining into my eyes. And then in an instant it was past, moving along the road, closer and closer to her. I could see it slow, and in the distance stop. The truck door opened and the shadow of her climbed in.

And in that moment she was gone.

ACKNOWLEDGEMENTS

Many thanks to those stalwart early supporters of this book—Marele Day, Siboney Saavedra-Duff, John Pitt, Elizabeth Maddox and everyone at the Northern Rivers Writers' Centre.

To Varuna, the Writers' House, for its generous support, and all the lovely writers I've met there over the years. Special thanks to the wondrous Peter Bishop, the most insightful and enthusiastic champion a writer could ever wish for.

To my earliest readers—Sarah Armstrong, Anna Sabadini, Kate King, Matt Hagan, Jacob Cole, Jan Smith, and Neil and Lyn Buhrich. Your thoughtful feedback was incredibly important.

For all your professional insights, thanks to Angie Carroll, Torben Wentrup and Vincent Duff.

Thanks also to Varda Shepherd, Simon Bender, Louise Nicholls, Danika Cottrell, Rose Anderson, Bradley McCann, Luke Wright, James Murray, Ricci Raso, Jahnavi Vinden-Clark, Penny Nelson, Niki Huang, Bridget Harding, Jane Camens, Helen Bums, Lisa Walker and Salvatore Calabro for friendship, inspiration and emotional support.

To Donica Bettanin and Jenny Darling for picking me out of the crowd and keeping me in the loop.

For the first edition, many thanks to the HarperCollins team, especially my editor Mary Rennie, who always offered such valuable counsel. For this beautiful second edition, big thanks to Matt Rubinstein and Ligature Publishing.

I'd also like to acknowledge both Bruce Springsteen and Lucinda Williams, without whose bodies of work this book would never have been written.

And, as always, a very special thanks to my family.

ABOUT THE AUTHOR

Jessie Cole is a writer. Her first novel, *Darkness on the Edge of Town*, was shortlisted for the 2013 ALS Gold Medal and longlisted for the Dobbie Literary Award. Her second novel, *Deeper Water*, was released in 2014 to critical acclaim. *Staying*, a memoir, was longlisted for the 2019 Colin Roderick Award and shortlisted for the Victorian Premier's Literary Award for Non-Fiction. Her latest memoir, *Desire, A Reckoning*, was released in 2022. She lives in northern New South Wales.